Project AIR FORCE

Supporting Expeditionary Aerospace Forces

ENGINE MAINTENANCE SYSTEMS EVALUATION (ENMASSE)

A USER'S GUIDE

Mahyar A. Amouzegar
Lionel A. Galway

Prepared for the
UNITED STATES AIR FORCE

RAND

The research reported here was sponsored by the United States Air Force under Contract F49642-01-C-0003. Further information may be obtained from the Strategic Planning Division, Directorate of Plans, Hq USAF.

Library of Congress Cataloging-in-Publication Data

Amouzegar, Mahyar A.
Supporting expeditionary aerospace forces : engine maintenance systems evaluation (EnMasse) : a user's guide / Mahyar A. Amouzegar, Lionel A. Galway.
p. cm.
"MR-1614."
Includes bibliographical references.
ISBN 0-8330-3285-2
1. Jet engines—United States—Maintenance and repair—Simulation methods. 2. United States. Air Force—Equipment—Maintenance and repair—Simulation methods. 3. Airplanes, Military—United States—Maintenance and repair—Simulation methods. I. Galway, Lionel A., 1950– II.Title.

UG1103 .A48 2003
623.7'06044—dc21

2002151079

Published 2003 by RAND
1700 Main Street, P.O. Box 2138, Santa Monica, CA 90407-2138
1200 South Hayes Street, Arlington, VA 22202-5050
201 North Craig Street, Suite 202, Pittsburgh, PA 15213-1516
RAND URL: http://www.rand.org/
To order RAND documents or to obtain additional information, contact Distribution Services: Telephone: (310) 451-7002;
Fax: (310) 451-6915; Email: order@rand.org

PREFACE

This report is a user's guide for the Engine Maintenance Systems Evaluation (EnMasse), a simulation model used in the analysis of alternative concepts for Jet Engine Intermediate Maintenance (JEIM). The result of that analysis is reported in a companion document, *Supporting Expeditionary Aerospace Forces: Alternatives for Jet Engine Intermediate Maintenance*, MR-1431-AF, 2002. This is ongoing research in support of emerging Air Force employment strategies associated with Expeditionary Aerospace Forces (EAFs). EAF concepts rely on the premise that rapidly adaptable, quickly deployable, immediately employable, and highly effective air and space force packages can serve as a credible substitute for permanent forward presence. Success of the EAF will, to a great extent, depend on the effectiveness and efficiency of the Agile Combat Support (ACS) system. This report is one of a series of RAND publications that address ACS issues in implementing the EAF. Others address planning, practices, policies, and technologies that can enhance the effectiveness of the EAF. Titles in this series include

- *Expanded Analysis of LANTIRN Options* (MR-1225-AF, 2001),
- *Alternatives for Jet Engine Intermediate Maintenance* (MR-1431-AF, 2002),
- *Flexbasing: Achieving Global Presence for Expeditionary Aerospace Forces* (MR-1113-AF, 2000),
- *An Analysis of F-15 Avionics Options* (MR-1174-AF, 2000)
- *New Agile Combat Support Postures* (MR-1075-AF, 2000),

- *A Concept for Evolving the Agile Combat Support/Mobility System of the Future* (MR-1179-AF, 2000), and
- *An Integrated Strategic Agile Combat Support Planning Framework* (MR-1056-AF, 1999).

The research addressed in this report was conducted in the Resource Management Program of Project AIR FORCE (PAF) as one element of a project entitled "Implementing an Effective Air Expeditionary Force." The project was sponsored by the Air Force Deputy Chief of Staff for Installations and Logistics (AF/IL), Air Combat Command's Director of Logistics (ACC/LG), and, in its early stages, jointly by the Air Force Deputy Chief of Staff for Plans and Operations (AF/XO). This report should be of interest to logisticians and operators in the Air Force concerned with implementing the EAF concept.

PROJECT AIR FORCE

Project AIR FORCE, a division of RAND, is the Air Force federally funded research and development center (FFRDC) for studies and analyses. It provides the Air Force with independent analyses of policy alternatives affecting the development, employment, combat readiness, and support of current and future aerospace forces. Research is being performed in four programs: Aerospace Force Development; Manpower, Personnel, and Training; Resource Management; and Strategy and Doctrine.

CONTENTS

FIGURES

TABLES

SUMMARY

This report is a user's guide for the Engine Maintenance Systems Evaluation (EnMasse). EnMasse is a simulation model based on Extend software and used in the analysis of alternative Jet Engine Intermediate Maintenance (JEIM) policies. The result of the policy analysis conducted using EnMasse is reported in a companion document, *Supporting Expeditionary Aerospace Forces: Alternatives for Jet Engine Intermediate Maintenance*, MR-1431-AF, 2002.

The goal of the analysis was to evaluate several alternatives for accomplishing JEIM support. Closely allied to maintenance policy are the maintenance structures within which these policies operate both in peace and war.

In terms of modeling and simulation, we are interested in the flow of entities (e.g., spare engines, personnel), the state of the system (e.g., engines not mission capable, spares inventory), and the processes (e.g., service time, sortie rates). EnMasse's structure is based on a set of hierarchical, functional blocks that generate and modify entities, processes, and attributes. These blocks represent Air Force home bases, flightlines, JEIM shops, module shops, test cells, forward support locations (FSLs), and forward operating locations (FOLs). This report is not a traditional user's guide in that it does not aim to give the user an exhaustive list of model inputs and outputs, but rather its goal is to allow the user to examine the model using the Graphical User Interface (GUI) and determine and modify functions from that vantage point.

In general, EnMasse is based on the following sequence of events: aircraft are flown from home bases and FOLs to meet peacetime (training) and wartime flying schedules, respectively. After each mission, the aircraft and their engines are inspected at the flightline, and in most cases they are fully operational within hours. However, when engines accumulate enough flying hours, or when unscheduled maintenance is required, they are removed from the planes and sent to a JEIM facility. They are then inspected, repaired, tested, and returned to service. The JEIM facility includes the JEIM shop, the module shop, and the assembly and test cell.

The first requirement for each model is the number and types of aircraft (e.g., F-15, F-16), and the number and ages of the installed engines. Both aircraft and engines are required to form fully mission capable (FMC) aircraft.

After each sortie, aircraft are sent to the *Flightline* block where they are inspected and maintained. Each aircraft that passes the inspection is sent back to the pool of available aircraft. Some aircraft require minor repairs, which are performed on the flightline. EnMasse also allows for scheduled and unscheduled maintenance. The number of engines pulled from the aircraft is a function of the age and the type of the engine. The detached engines are tagged according to the removal type and sent to the JEIM shop. Aircraft tagged as not mission capable (NMC) are sent to the *Spare Engines Analysis* block where they are queued for the next available engine. If serviceable spares are available, these aircraft are put back into service immediately. Otherwise, they await the arrival of engines from the *Assembly and Test Cell* block.

The *JEIM Shop* block requires two inputs from the user: the initial number of labor units and the number of rails (i.e., the JEIM capacity). Engines are queued in two parallel lines, the first for engines that require parts that are not available and the other for engines that await maintenance. The modular engines that have been processed by the JEIM shop are sent to the *Module Shop* block.

Engines that enter the Module shop are separated into five modules: fans, core, low-pressure turbine (LPT), augmentor, and gearbox. Engines that leave the module shop are sent to the *Assembly and Test Cell* block. In this block, engines are queued for assembly, the test cell and the final inspection. After assembly and test cell, engines are sent to the *Spare Engines Analysis* block. In this block, the FMC engines are pooled with the other spares (including the war reserve engines) and queued for installation on the aircraft. The FMC aircraft leave this block to join the pool of other aircraft, and the whole cycle starts again.

ACKNOWLEDGMENTS

The research documented in this report could not have been done without the active cooperation and help of a large number of Air Force people. In these acknowledgments we list their positions when the study was done.

Much of the RAND work on Expeditionary Aerospace Force support has been done under the sponsorship of several different Deputy Chiefs of Staff for Installations and Logistics (AF/IL). Much of this work was done when Lt Gen John W. Handy held this position. Our point of contact with AF/IL was primarily Susan O'Neal (AF/ILX), who together with her staff provided much useful guidance, information, and contacts with other Air Force organizations. As described in the report, this particular project was co-sponsored by ACC/LG Maj Gen Dennis Haines. We are particularly indebted to ACC/LGSP for its help and involvement as our point of contact with ACC, especially Col Stanley Stevens, Lt Col John Cooper, CMSgt Hank Houtman, CMSgt Michael Kinser, and ACC Command Engine Manager Tom Smith. Data were also provided by ACC/LGP, under the direction of Ed Merry.

At San Antonio Air Logistics Center, we would like to thank Robert May (SA-ALC/LR) and his staff, particularly Melissa Tinscher and Chris Szczepan, for information on engine requirements. We also benefited from discussions with SA-ALC/LPF, Colonel Doumit and his staff, especially Colonel McMahon, Greg Hall, Bruce Eberhard, and David Crowley.

For access to and patient help with data from the Comprehensive Engine Management System, we are grateful to Charlie Osborn, Phil

Garrity, Jim Blain, David Addison, and Walt Cooper. For help with access to and interpretation of data from the Reliability and Maintainability Information System (REMIS), we appreciate the help of Richard Enz and Thomas Recktenwalt.

We had extensive help from a number of JEIM shops and their senior NCOs. These include SMSgt Kasprak at the 1st Fighter Wing (Lang ley AFB, Va.), SMSgt McDyer at the 20th Fighter Wing (Shaw AFB, S.C.), SMSgt Dotten 347th Fighter Wing (Moody AFB, Ga.), SMSgt Travis at the 366th Fighter Wing (Mountain Home AFB, Idaho), CMSgt Mackey at the 48th Fighter Wing (RAF Lakenheath, UK), SMSgt Gasper at the 31st Fighter Wing (Aviano AB, Italy), and CMSgt Holas at the 52nd Fighter Wing (Spangdahlem AB, Germany). All of these very busy people and their staffs graciously organized tours of their shops and meetings with their supervisors, collected data and information on their operations, and fielded clarification questions after our visits. This project also drew on a discussion with CMSgt Hauck at USAFE headquarters on engine support during Operation Noble Anvil.

Finally, we appreciate the hospitality of the Air Force Research Laboratory at Wright-Patterson AFB, Ohio, for a day of discussions on new propulsion technologies.

As always, we benefited greatly during our project from the comments and constructive criticism of many RAND colleagues, including Laura Baldwin, Richard Moore, C. Robert Roll, and Hyman Shulman. We owe special thanks to Robert Tripp, the project leader for all of the EAF support tasks, for his guidance and support and to Louis Miller for his insightful comments on building a robust simulation model. Marygail Brauner provided detailed reviews of drafts of this work that substantially improved the quality and the presentation. Frances Teague and Kristin Leuschner patiently and promptly did an excellent job editing the document.

Our modeling in this project was done with Extend simulation software. We had very good technical support from author Bob Diamond and his people at Imagine That, Inc., in San Jose, Calif.

ACRONYMS

ACC	Air Combat Command
ACS	Agile Combat Support
AEF	Aerospace Expeditionary Force
AETC	Air Education and Training Command
AFB	Air Force Base
AFLMA	Air Force Logistics Management Agency
AFMC	Air Force Materiel Command
ALC	Air Logistics Center
ANG	Air National Guard
AWM	Awaiting maintenance
AWP	Awaiting parts
CAMS	Core Automated Maintenance System
CEMS	Comprehensive Engine Management System
CIRF	Centralized Intermediate Repair Facility
CONUS	Continental United States
CSL	CONUS support location
EAF	Expeditionary Aerospace Force

EMB	Engine Management Branch
EnMasse	Engine Maintenance Systems Evaluation
ENMCM	Engine not mission capable because of maintenance
ENMCS	Engine not mission capable because of supply
FCFS	First come, first served
FIFO	First in, first out
FMC	Fully mission capable
FOD	Foreign object damage
FOL	Forward operating location
FSL	Forward support location
GUI	Graphical User Interface
JEIM	Jet Engine Intermediate Maintenance
LANTIRN	Low-Altitude Navigation and Targeting Infrared for Night
LPT	Low-pressure turbine
MAJCOM	Major Command
MDS	Mission Design Series
MTW	Major theater war
NMC	Not mission capable
OSD	Office of the Secretary of Defense
PAA	Primary Aircraft Authorized
PRS	Propulsion Requirements System
REMIS	Reliability and Maintainability Information System

SER	Scheduled engine removal (rate)
SSC	Smaller-scale contingency
SWA	Southwest Asia
TAMS	Tactical Aircraft Maintenance Specialists
TC	Test Cell
TCTO	Time Change Technical Order
UER	Unscheduled engine removal (rate)
USAFE	U.S. Air Forces in Europe
UTC	Unit Type Code
WRE	War reserve engine

Chapter One

INTRODUCTION

This report provides a detailed guide to Engine Maintenance Systems Evaluation (EnMasse), a simulation model developed by RAND to analyze jet engine intermediate maintenance alternatives for the U.S. Air Force.

This analysis was prompted by the ongoing reorganization of the Air Force into an Expeditionary Aerospace Force (EAF). The main objective of this reorganization is to replace the forward presence of air power with a force that can deploy quickly from the continental United States (CONUS) in response to a crisis, commence operations immediately on arrival, and sustain those operations as needed. To support the expeditionary force, such support processes as munitions, fuels, and maintenance also need to be transformed. EAF requires a combat support system capable of supporting an expanded range of operations, including major theater wars and smaller-scale contingencies (SSCs), which could take place in any of a number of different locations.[1]

Since 1997, RAND has conducted a series of studies for the Air Force to understand how combat support can be adapted for expeditionary operations.[2] The most important finding of the work to date is that the Air Force's original goal of deploying a complete package of combat aircraft and support within 48 hours to an unprepared ("bare

---

[1] For a more complete description of the EAF concept and its history, see Davis (1998) and Ryan (1998).

[2] See Galway et al. (2000); Tripp et al. (2000); Feinberg et al. (2001); and Peltz et al. (2000).

base") forward operating location (FOL) cannot be met with current support processes. The timeline can be met only with judicious prepositioning of materiel at the FOLs and the establishment of forward support locations (FSLs) for storage and maintenance of selected commodities. Complete deployed support can be provided for fighter units from CONUS only by accepting a timeline on the order of a week or more.

JET ENGINE INTERMEDIATE MAINTENANCE UNDER EXPEDITIONARY OPERATIONS

The analysis of support strategies for the EAF has subsequently been extended to other critical processes to determine where they should be located. One of these critical processes is Jet Engine Intermediate Maintenance (JEIM), which provides combat units with extensive repair of jet engines. The JEIM facility consists of several components, including the JEIM shop, the module shop, and the assembly and test cell. JEIM is one of three levels of maintenance used by the Air Force to repair jet engines, especially those powering fighter aircraft:

- **Flightline maintenance** consists mostly of inspections, diagnostics, engine removals, and some quick repairs that do not involve engine teardown.
- **Intermediate maintenance** at the JEIM facility includes disassembly of the engines; substantial repairs to such parts as fans, low-pressure turbines (LPTs), and afterburners; and engine test cell runs.
- **Depot maintenance** involves the complete teardown and refurbishment of any repairable part in an engine. The rebuilding of an engine at the depot allows the engine's use parameters (flight time, cycles, etc.) effectively to be reset at zero.

Traditionally, the JEIM has been located at the operating base with the aircraft and under the overall command of the operational commander.[3] This practice stemmed from the long-held concept that the operational commander should have control of all of the resources needed to generate required sorties and that the unit should be relatively self-sufficient in combat and combat-support capability for a period of weeks. This policy was reinforced by the planning for major wars in Europe and Korea: A unit would be moved to existing bases in theater in preparation for immediate action and could expect little resupply during the first few weeks of combat. Under traditional planning for wing deployment, therefore, the JEIM is prepared to move along with the rest of the wing support, although not with the combat units themselves, who will use spares to replace engines until the JEIM arrives and is up and running.

In recent years, the question of whether JEIM operations should be centralized has been the subject of frequent discussion in the Air Force engine community. Many factors favored centralization, including the increased complexity of engines and the large investment required for repair facilities. Other factors worked against centralization, particularly the fact that, unlike such other commodities as avionics components, engines are heavy and bulky and thus require special packing to ship. Over the years, the Air Force has experienced a pattern of alternation between the partial centralization of JEIM operations—in certain regions and for certain engine types—and the subsequent restoration of JEIM facilities to operating units.

The requirements associated with expeditionary operations—including the ability to move quickly, the need to keep initial transportation requirements down—have raised new questions about the policy of locating the JEIM facility at the operating base. Our current research aims to provide insights into this issue by determining whether JEIM support can best be provided from decentralized shops at the supported bases or from a centralized, off-base facility. The results of this analysis are reported in a companion document, *Supporting Expeditionary Aerospace Forces: Alternatives for Jet Engine Intermediate Maintenance,* MR-1431-AF, 2002.

---

[3]For very reliable engines, especially those in transport aircraft, which spend large amounts of time away from their home bases, the JEIM has sometimes been located in a regional or "Queen Bee" facility.

DEVELOPMENT OF SIMULATION MODEL

This report focuses on the suite of simulation models we developed to understand and evaluate the support alternatives for the JEIM. These models, referred to collectively as EnMasse, were created using a system and process modeling software package known as Extend.[4] EnMasse offers dynamic modeling capabilities that allow the user to create a realistic simulation of the jet engine repair system. It simulates the interaction among the components of the maintenance system, while incorporating the random variations or uncertainties typical of a dynamic system. Using EnMasse, we could analyze a number of possible support configurations for the JEIM, involving various combinations of centralized and decentralized locations. The simulation models allowed us to compare several alternatives for maintenance support across different scenarios.

This report focuses on the development, use, and modification of the EnMasse simulation model in analyzing maintenance alternatives. Our objective is to provide a basic understanding of the key features and capabilities of EnMasse. Although the report includes some information about the use of Extend, it is not intended as a software user's manual. Readers interested in learning more about the capabilities and functioning of Extend should refer to the Extend user's manual.[5]

REASONS FOR SELECTING MODELING AS THE METHOD OF ANALYSIS

In this section we describe our reasons for developing a fairly complex simulation model as the primary means of analyzing alternatives for JEIM support. The simulation model provides several advantages for analyzing and comparing jet engine maintenance support options.

As stated earlier, our aim in this project was to compare several alternatives for locating the JEIM. These alternatives included full centralization in peace and war, as well as several hybrid systems

---

[4]Extend was created by Imagine That, Inc.

[5]For more information, see www.imaginethat.com.

(e.g., decentralized in peace but centralized in the theater of operations). The alternatives were to be compared using several performance metrics and potentially several different scenarios as well. We could have used two main approaches for this analysis:

- Use data from the previous history of centralization attempts to determine whether centralization will work.
- Develop a model of the JEIM and supporting systems, such as transportation, and evaluate the alternatives within the model.

In our view, the history of centralization was of limited use in assessing JEIM alternatives. In many of the historical centralization efforts—both successful and unsuccessful—decisions about location were driven by external constraints, which may not apply in general situations. Moreover, limited data were available on pre- and post-centralization performance, and no information was available on any of the major centralized facilities during a conflict. This is not surprising because almost no centralized facility has supported a conflict as the major source of repair. We also wanted to examine the effects of centralizing intermediate repair for engines that had never had centralized repair (e.g., the F100-229), to look at full centralization of engines that had partially centralized repair (e.g., the TF-34), and to look at engines for which centralization had failed (e.g., the F100-220).

For all these reasons, we turned to modeling as our primary tool for the analysis. In developing the model, however, we drew on the history of past centralization efforts in selecting the alternatives to be analyzed and understanding some of the key factors that have traditionally caused problems in centralized repair.

ADVANTAGES OF A SIMULATION MODEL

Our next step was to determine which kind of model would best suit our objectives. One option we considered was an "expected value" model, which uses the means of stochastic quantities (e.g., transportation times) in deterministic formulas, which are supplemented by uncertainty computations, such as confidence limits. Much previous RAND work in support of the EAF has used this type of model.[6]

Another option, and the one we chose, was a simulation model. Unlike expected-value models, a simulation model is capable of directly incorporating aspects of uncertainty. For our purposes, a simulation seemed advantageous for several reasons:

- **Accommodation of dynamic metrics.** The metrics in which we were potentially interested—sorties missed, current spare levels, queue sizes at key shop points, etc.—are inherently dynamic, and we wanted to see the value of key metrics day by day. For example, during conflict situations, sortie requirements may change. Under such circumstances, a force may miss only 5 percent of required sorties, but there is a big difference in performance if that 5 percent is concentrated during the first few days of a war rather than at the end of a conflict.
- **Flexibility in setting time dimensions for the analysis.** Management decisions about engine deployment and repair are regularly based on the time characteristics of individual engines. For example, when a unit is deploying for operations away from home, the propulsion flight supervisors try to select those engines with the most time remaining until major inspections or other work.
- **The ability to include engine "demographics" in the analysis.** Demographics refer to the age distribution in terms of such parameters as cumulative flying hours. Engine demographics drive the inspection and removal of many critical components of the engine and are key to the performance of the repair system. Depot repair, as noted above, usually "zero-times" the engine. The distribution of engine ages at a particular point is an important determinant of JEIM (and depot) workload. Conversely, modifying workload can manipulate the age distribution.
- **Variability in setting repair "modes" in order to analyze their impacts.** Some engines have several repair "modes," depending on whether an engine removal is scheduled (for an inspection or

---

[6]See Feinberg et al. (2001) and Peltz et al. (2000).

to change a part that has reached a specific age) or unscheduled (due to a malfunction of some type). In addition, for some types of engines, such as the TF-34, the engine repair can either be a quick-turnaround repair or a more complete disassembly. The proportion of each type of repair can have different effects on the internal work flow of the shop.

- **The ability to analyze potential transportation options at a relatively high level of detail and to incorporate other transportation variables in addition to transportation times.** These variables include limited transportation capacity, transportation schedules, and such options as waiting until two engines need to be shipped to minimize shipping costs.

In addition to these considerations, some initial experimentation indicated that current Graphical User Interface–based[7] simulation packages could indeed provide us with a simulation that ran in reasonable times when simulating repair operations for current fighter engine fleets.

Simulation Modeling

All of these considerations led to our decision to build a simulation model. Although the model required a substantial investment in initial effort, the result was a flexible tool that could be used for this and future investigations.

Simulation models, such as EnMasse, attempt to predict the behavior of the system under investigation by replicating and analyzing the interaction among its components. In the past, one had to compromise between choosing a model that provided a realistic replica of the actual situation and one whose mathematical analysis was tractable. With the advent of faster computers and increased memory, we can develop a more realistic reflection of reality without compromising on mathematical rigor.

By expressing the interactions among the components of the system as mathematical relationships, we can gather information in much

[7]GUI enables a wider access to the power of the digital computer.

the same way as if we were observing the real system (subject, of course, to the simplifications built into the model). Simulation thus allows greater flexibility in representing complex systems that are normally difficult to analyze by standard mathematical models. We must keep in mind, however, that a model by definition is not the real world, but its reflection. No matter how hard we try, we will miss many nuances of the real world. In the end, we must make some compromises to get reasonable results. We can reduce the effect of such compromises, as we have done in the main study, by additional analysis of the problem.[8]

ORGANIZATION OF THIS REPORT

Traditionally, documents that describe themselves as "user's guides" for simulation programs have a generic structure. They begin with a description of the real-world process being modeled; explain a bit about key algorithms in the model (random number distributions, queue disciplines), especially where they embody assumptions and approximations; and then provide a detailed and exhaustive catalog of model inputs and outputs. In some cases, this is supplemented by program flowcharts and actual code listings, but for large, complex models inclusion of this latter material is quite rare.[9]

Because we use a state-of-the-art, GUI modeling language, this report is somewhat different. First, the model itself is largely isomorphic to the real-world system: The shops in the JEIM are identifiable entities in the model, as is the transportation system and the flightline. In most cases, a user can therefore "read the code" directly by knowing what the real-world system looks like. By implication, it is unnecessary for us to document each instance when a random number generator is used or what the discipline for a particular queue is. Similarly the flow of engines and information into and out of each block of the simulation can be determined from the model itself. In short, "reading the code" is now feasible as a way of understanding a particular model.

---

[8]For the result of this analysis, see Amouzegar, Galway, and Geller (2002).

[9]For an outstanding example, see Isaacson and Boren (1993).

Our first goal is therefore to describe in some detail the real-world system we are modeling, namely the operation of the JEIM. We then describe how we have represented the functions of the JEIM, the flightline, the transportation system, and other key elements as Extend structures (queues, decisions, etc.). This is done at a fairly high level with enough detail for a user to understand our major assumptions and approximations. The ultimate goal is that a user can take our Extend blocks and combine them into a structure for JEIMs that will allow its performance to be simulated and hence evaluated. In many cases, that will require some modifications, but our description should allow an Extend user to find the relevant blocks and modify them. Unlike the case with traditional guides, we do not anticipate that a user would try to run an EnMasse model by using the information here alone. Instead, this would form the basis for an understanding of the code. This implies that a reader would have some basic familiarity with the elements of Extend (or has a manual available). For example, we often comment that a particular block can be edited to change parameters, etc. This is an Extend operation in which the user double-clicks on a block and opens a dialogue block, which contains the parameters and allows them to be modified as with any text.

The remainder of the report is organized as follows. Chapter Two provides a detailed description of JEIM maintenance support options and an overview of the simulation. Chapter Three focuses on the structure of the model. Chapter Four describes the components of the EnMasse library and the steps in building various functioning models. It also illustrates how the existing library can be modified to accommodate alternative models.

Appendix A presents a sample run of the model, and Appendix B illustrates detailed diagrams of the blocks in the EnMasse library.

Chapter Two

SIMULATION OF ENGINE MAINTENANCE SYSTEMS

This chapter provides an overview of the simulation model. First, we sketch out the overall requirements of the analytical model by describing the key components of the flightline and JEIM activities to be replicated in the simulation. We then take a look at the model's functions and describe the maintenance alternatives assessed with the model and the metrics used to evaluate the results.

FLIGHTLINE AND JEIM MAINTENANCE

A brief review of the key components and functions of the flightline and JEIM operations will provide a foundation for the discussion of the simulation.

As noted in the Chapter One, the flightline provides inspections, diagnostics, and quick repairs, while the JEIM is responsible for off-equipment engine maintenance that does not involve complete teardown and rebuilding. In many instances, the JEIM assists the flightline as well.

Flightline maintenance includes servicing, repairs, cycle recording, and tracking, which are coordinated with the Engine Management Branch (EMB) and JEIM. On the flightline, installed aircraft engines are serviced daily by the Tactical Aircraft Maintenance Specialists (TAMS). Flightline activities include servicing the oil, inspecting the chip detectors, and entering the intakes and augmentor to inspect for foreign object damage (FOD) and external engine damage. In addition, engine cycles are recorded in the Comprehensive Engine Management System (CEMS) database. CEMS enables the EMB to monitor usage of engines and modules (when used) to determine the need for inspections and Time Change Technical Orders (TCTOs). The flightline also performs all engine removals and installations. After the flightline removes an engine for maintenance at the JEIM, it sometimes performs sheet-metal work on the engine bay and replaces some of the hydraulic lines and cables in the aircraft engine bay that have been damaged due to chafing, cracks, or heat.

The JEIM is responsible for both scheduled and unscheduled off-equipment engine maintenance. Scheduled maintenance includes module time changes, TCTOs, and other inspections and repairs. Unscheduled maintenance consists primarily of performance-related problems that either cannot be corrected by the flightline or are beyond their capabilities per Technical Order. For unscheduled maintenance, the JEIM shop often performs a preliminary test cell run to troubleshoot the engine and identify potential problems. The JEIM is capable of replacing any module in a modular engine and also repairs some of the modules while sending others to the depot. It is also responsible for packing engines for transportation.

The JEIM operates the engine test cell facility and functions. As part of this, the JEIM transports engines, hooks up cables and fuel lines, conducts pre- and postrun engine inspections, disconnects cables and fuel lines, and transports the engines to the JEIM shop.

In many cases, the JEIM is also a source of expertise to back up the flightline and provide quick response repair or cannibalizing key parts as needed. The collocation of JEIM with fighter squadrons has resulted in a slight blurring of the functions of the two lower repair levels. However, only personnel from the wing JEIM are authorized to sign off on JEIM-level repair work for the wing's engines.

The JEIM is staffed by the propulsion flight (usually a part of the Component Repair Squadron). This organization is quite large (100–150 people for a fighter wing) and occupies an industrial space equipped with five or more work bays of 1,500 square feet each, an overhead crane, supply storage, backshops for specialized repair activities, and a test cell. The test cell is typically located off site in a "hush house," where a fully assembled engine can be run at full power for testing purposes.

The general flow of JEIM work is as follows:

- receive engine from the flightline;
- perform TCTO and time change check;
- perform CEMS history check;
- create job in Core Automated Maintenance System (CAMS);
- assign engine to a crew;
- determine required repairs;
- decide on complete or partial disassembly;
- conduct other inspections;
- conduct teardown;
- perform JEIM repair and maintenance;
- perform module work, if needed, at module shop;
- assemble engine;
- send to test cell (hush house); and
- conduct final inspection.

ENGINE MAINTENANCE SIMULATION

We now provide an overview of the EnMasse simulation. The first step in constructing the simulation model was to express the real system in terms of its key *events*. An event is defined as a time at which changes in the character of the system take place. For example, an arrival of an NMC engine to the JEIM shop is an event in the engine maintenance system.

Each model in EnMasse is based on a basic sequence of events. First, aircraft are flown from bases and FOLs to meet peacetime (training) and wartime flying schedules, respectively. After each mission, the aircraft and their engines are inspected at the flightline and in most cases are fully operational within hours. However, when engines accumulate enough flying hours,[1] or when unscheduled mainte-

[1]The model keeps track of engine serial numbers and aircraft tail numbers throughout the simulation.

nance is required, they are removed from the planes and sent to a JEIM facility, which includes the JEIM shop, the module shop, and the assembly and test cell. After arrival at the JEIM shop, engines are inspected, repaired, tested, and returned to service.

As indicated by Figure 2.1, EnMasse is a closed-loop, discrete-event simulation model. A closed-loop model implies that entities (e.g., engines, personnel) never enter or leave the model although the state of the system (the occurrence of events) changes at random times. While engines, aircraft, or people may move from bases to FOLs or centralized repair facilities, the total number of such entities in the system is fixed. This closed-loop property has important implications for the dynamic interactions between repair and usage. For example, if an engine shop can fix more engines, then fewer aircraft will have holes. As a result, fewer sorties will be missed because of engine unavailability, and therefore a greater number of engines will be used. However, the increase in utilization could increase the number of engines failing, which in turn would put pressure on the engine shop and ultimately reduce its production rate. The decrease in production would have an effect (excluding the effect of spares) on flying hours, which in turn could reduce the number of engines failing. EnMasse can capture the dynamic nature of such relationships within the maintenance system.

MAINTENANCE ALTERNATIVES AND METRICS

Using EnMasse, we analyzed a number of possible support configurations for the JEIM that involved various combinations of centralized and decentralized locations. Centralized maintenance structures include FSLs, while decentralized locations include home base support and maintenance at FOLs. Each structure was assessed under both wartime and peacetime scenarios.

Here we describe in detail the specific JEIM alternatives we evaluate in this analysis:[2]

---

[2]We label each alternative in terms of "peacetime repair–wartime repair." For example, the decentralized-deployed case implies a decentralized mode of repair during peacetime (and for nonengaged forces) at home units and deployed JEIM shops at the FOLs.

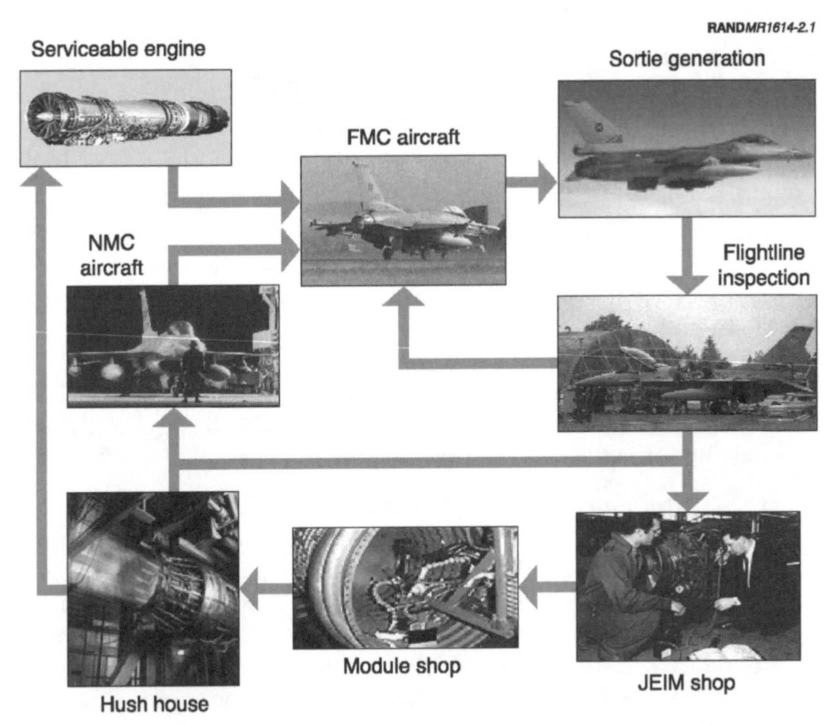

SOURCE: Photo (lower left) courtesy of Pratt & Whitney, A United Technologies Company. All other photos courtesy of the U.S. Air Force at http://www.af.mil.

Figure 2.1—Closed-Loop Maintenance Flow

- ***Decentralized-Deployed.*** In this alternative (which is the current plan for deployed engine support), peacetime maintenance is provided by JEIMs at each base. When part of a unit is deployed, part of that unit's JEIM deploys to the appropriate FOL to form a deployed JEIM. According to current plans, the JEIM deploys by day 30 of the war and begins working immediately, but the test cell is not ready to test repaired engines until day 60.[3] The trans-

---

[3]This limitation stems from the requirement that the test cell foundation must be strong enough to resist the thrust of modern fighter engines at full military power

portation requirement for this alternative is that needed to deploy the JEIM itself.

- ***Decentralized–No Deployment.*** As with the previous alternative, each of the peacetime bases has its own JEIM, but in this case the home JEIM supports any deployed forces from its unit as well.[4] The home JEIM is sized to have the resources to support both peacetime and wartime flying.
- ***Decentralized-FSL.*** As with the previous two alternatives, each peacetime base has its own JEIM, but, when the units deploy, some of the JEIM personnel (but not their equipment) deploy to a single FSL in-theater from which all deployed units are supported. We assume that the FSL is "lukewarm"—i.e., it is ready to begin operations as soon as the JEIM personnel arrive. In this case, no additional delay occurs for the test cell setup but there may be some delay for the arrival of the personnel.
- ***CONUS Support Location–FSL (CSL-FSL).*** In this alternative, all units are supported in peacetime by a CSL, which deploys personnel to an FSL in-theater when conflict occurs. In peacetime, the CSL is staffed with the sum of the rail teams[5] needed for deployment and those required to keep the nonengaged forces flying. (Note that for deployed forces, this alternative and the previous one are indistinguishable because the repair structure in-theater is identical.)
- ***CSL.*** In this last alternative, all units everywhere are supported by a single CSL both during peacetime and during deployment.

During the simulations, we evaluated each of these alternatives using three broad metrics. The first is performance: Does the alternative provide the required support for operational flying? In peacetime, this means maintaining the requisite flying hours for pilot training; in wartime, it means meeting the required number of sorties day by

(afterburner—about 29,000 pounds of thrust for the F100-229). The foundation is a concrete slab, which must set for 30 days after pouring.

[4]Note that some units use this method today to support their deployments to Operations Northern Watch and Southern Watch (enforcement of Iraqi no-fly zones).

[5]The engines are mounted on structures called "rails" for repair. A rail team is defined as the minimum number of personnel needed to work on an engine in a two-shift day.

day. The second metric is resources: What does the alternative require to provide adequate performance? For jet engines, one of the key resources is spare engines, which can provide a hedge against uncertainties. Other resources are personnel and transportation costs, and the evaluation provides an indication of the trade-off between these two. The third metric is uncertainty: How well does the alternative respond to unforeseen events? For this metric, we evaluate how robust the alternatives are to changes in the engine removal rate. EnMasse allowed us to compare alternatives in all three areas.

DATA SOURCES

Many of the inputs to the model were provided by analysis of data drawn from CEMS, Reliability and Maintainability Management Information System (REMIS), which rolls up data from the base-level CAMS, as well as data in both electronic and paper form provided by the units we visited.

The CEMS data provided information on total repair time for individual engines, engine NMC because of supply (ENMCS) times, and transportation times for such engines as the TF-34 for which Shaw AFB, S.C., provides JEIM repairs for some operational bases. REMIS provided a check on the CEMS data for overall engine repair and provided repair data for module work. However, neither could easily provide information linking module work to specific engines. CEMS started tracking module repair recently, and data series sufficient for analysis will be available in a couple of years. REMIS has space for the engine serial number in the module repair records, but the field is seldom used. We sought overall counts from REMIS of module repairs per engine inducted into the JEIM, but these were deemed unreliable because it was difficult to distinguish between scheduled and unscheduled work (many jobs are a mix, and the job is often coded as unscheduled work when it is started). For these reasons, module repair is an area of our modeling that requires more work.

The following chapter examines the structure of the model in detail.

Chapter Three

STRUCTURE OF THE MODEL

In this chapter, we demonstrate how to set up and run EnMasse using the FSL model for the JEIM shop supporting F-16 and F-15 aircraft carrying F100-229 engines.[1] EnMasse is a library of several modules that can be combined to develop scenario-specific engine maintenance models. Each module may receive data as an internal parameter (from user input and default settings) or from the output of another module. Some of the internal parameters are a function of types of engines or aircraft being simulated, while others, such as the number of aircraft, number of spares, etc., depend on the scenario being tested and are set by the user.[2]

Figure 3.1[3] illustrates the required user's input for a model in which units are supported at home base by their own JEIM and during deployment by an FSL in-theater. The FSL supports four bases: two F-16s (bases 1 and 2) and two F-15s (bases 3 and 4). The user must

[1]The steps for all the other scenarios are very similar to the FSL model. However, any differences will be highlighted below.

[2]The specific values were those used in the substantive study referenced above (Amouzegar, Galway, and Geller, 2002). Detailed explanations of how these values were derived can be found in that reference.

[3]Figure 3.1 is an example of an Extend *notebook*. In Extend, each instance of a block contains the parameters needed to control that block—e.g., parameters for a random number generator, service times for queues. This is convenient when programming but inconvenient for changing parameters across the model: a user would have to access each individual block and make detailed parameter changes. Extend notebooks link a single "control panel" to all blocks of interest and allow all relevant parameters to be changed from a single location. In this figure, the parameters illustrated are those that were of interest to the substantive study. Another user might want to modify the notebook to change other parameters.

decide on the number of aircraft and engines (e.g., 18 F-16s in each of bases 1 and 2, and 18 and 48 F-15s in bases 3 and 4, respectively), the sizes of the home JEIM (two units at base 1, one unit at base 2, three units at base 3, and seven units at base 4) and the module shop (two, one, four, and eight at bases 1–4, respectively), and the number of serviceable spares available at each unit (four, ten, twelve, and 24 at each respective base). Deployment inputs include three time entries, which are calculated based on a simulation start on day 1: when aircraft are deployed (day 360 for each unit),[4] when operations should commence (day 365 at both FOLs), and when they should terminate (day 465). Other deployment inputs include the number of aircraft needed from each unit (about two-thirds) as well as the time (day 360) and the number of spare engines (one, three, three, and eight, respectively) and labor units[5] (one, zero, one, and three from each JEIM shop and two, zero, two, and two from each module shop to the FSL).

The user must also decide on the number of prepositioned assets (two units at the JEIM shop and four at the MOD shop) and, finally, the transportation time from FOLs to the FSL (two days, one way).

Once the required user inputs have been entered, the next step is to decide on the duration of the simulation and the number of runs. The standard Extend pull-down menu can be used for this task. It can be located by clicking on the *Run* menu and then selecting *Simulation Setup*. Figure 3.2 illustrates the *Simulation Setup* menu. The simulation time should be entered in *days* (two years was used for the F100 engines).

If the user is satisfied with the model setup, no other input is required at this point, and EnMasse can start the simulation.

The rest of this chapter explains the hierarchical structure of the model and describes the inner workings of some of the modules, including special settings for running different maintenance options.

---

[4]Waiting for a year of peacetime operation allows the model to reach steady-state peacetime operation.

[5]The labor unit is a "rail team," a three- or four-person team that works a single shift daily on a single engine.

RANDMR1614-3.1

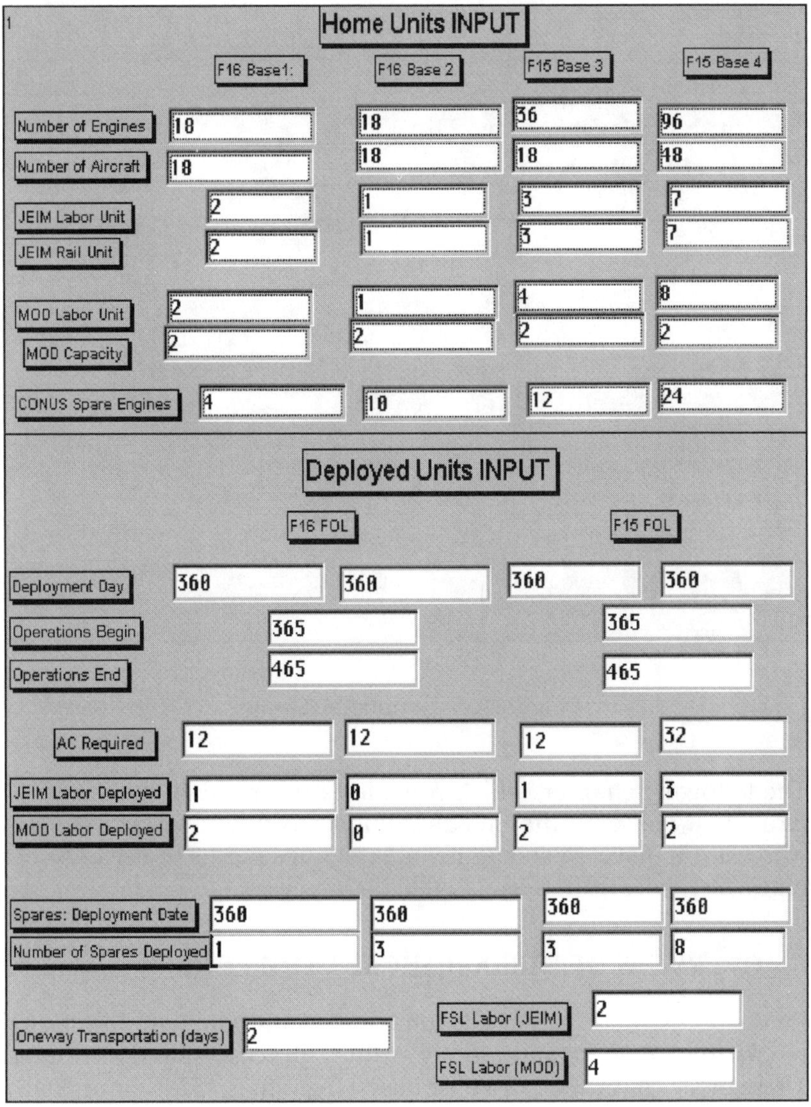

1

Home Units INPUT

	F16 Base1:	F16 Base 2	F15 Base 3	F15 Base 4
Number of Engines	18	18	36	96
Number of Aircraft	18	18	18	48
JEIM Labor Unit	2	1	3	7
JEIM Rail Unit	2	1	3	7
MOD Labor Unit	2	1	4	8
MOD Capacity	2	2	2	2
CONUS Spare Engines	4	10	12	24

Deployed Units INPUT

	F16 FOL		F15 FOL	
Deployment Day	360	360	360	360
Operations Begin	365		365	
Operations End	465		465	
AC Required	12	12	12	32
JEIM Labor Deployed	1	0	1	3
MOD Labor Deployed	2	0	2	2
Spares: Deployment Date	360	360	360	360
Number of Spares Deployed	1	3	3	8

Oneway Transportation (days) 2

FSL Labor (JEIM) 2

FSL Labor (MOD) 4

Figure 3.1—User Interface for an FSL Model

RANDMR1614-3.2

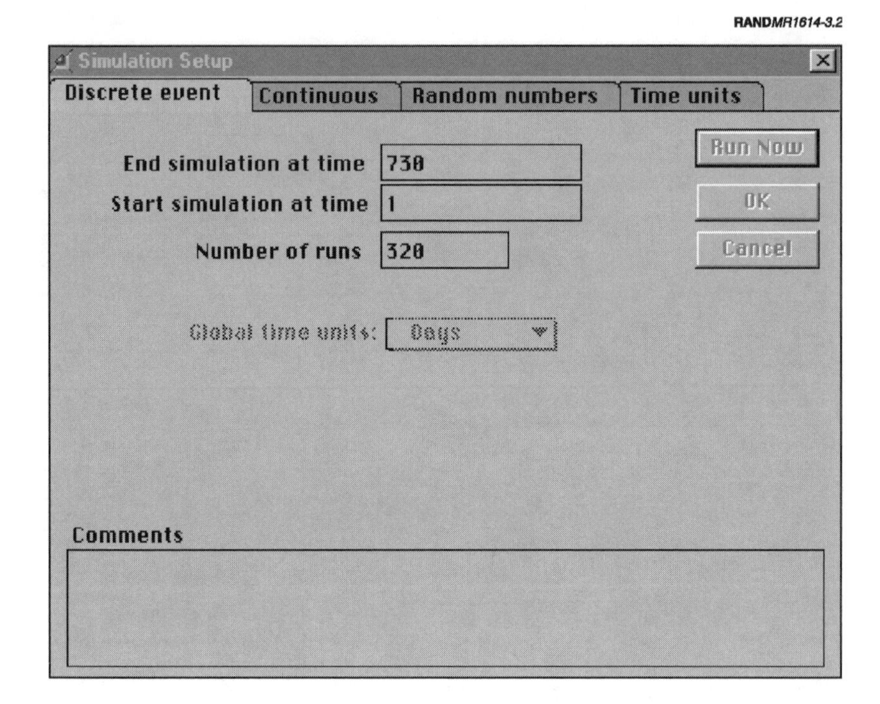

Figure 3.2—Simulation Setup

The following chapter, which provides a more detailed view of the individual blocks in the EnMasse library, is intended for the users who want to develop an engine maintenance model or use different types of engines from those used in our analysis.

AN OVERVIEW OF HIERARCHICAL MODELING

EnMasse's structure is based on a set of hierarchical, functional blocks that generate and modify entities, processes, and attributes. These blocks represent Air Force bases, flightlines, JEIM shops, FOLs, etc. The blocks are connected by "pipes," which transmit resources (such as engines, aircraft, and personnel) and information (such as engine identity and time) between the blocks, which can both generate and modify these entities. The complete process consists of a series of tasks and queues with each task requiring such resources as parts, personnel, and equipment.

The top level of the hierarchy provides the broadest view of contents of a model. In this chapter, we will examine the top-level hierarchy for an FSL model (Figure 3.3). For the second level of the hierarchy, we will look at an F-15 block that includes two home bases and an FOL (Figure 3.4) as well as a JEIM shop in an FSL (Figure 3.7). For the third level, we will look at an F-15 home base (Figure 3.5) and an FOL (Figure 3.6).

A Note on Reading the Figures

For the purposes of this report, we have simplified the representation of the EnMasse computer display in the figures shown in Chapters Three and Four. In these figures, we display the blocks that constitute the main process flow for each level of the model. However, we do not show all of the automated system inputs and other blocks that would be displayed on the computer screen during the running of EnMasse. We felt that a more streamlined design for the figures would be the most appropriate means of illustrating the structure and uses of the model. To review a complete set of EnMasse displays for all figures in this report, refer to Appendix B.

Inside the Upper Hierarchy

Figure 3.3 illustrates the upper hierarchy for a model using the decentralized-FSL scenario. Under this scenario, JEIM support would be decentralized during peacetime and provided from a centrally located FSL during a conflict. The upper hierarchy contains F-16 and F-15 blocks (labeled *F-16 World* and *F-15 World,* respectively), an *FSL* block that supports engines from engaged aircraft, and a transportation block that is used to simulate the transportation of engines from the FOLs to the FSL. This model uses ground transportation, although the block could easily be modified to include air or sea transport as well.

Starting from the left side of the figure, the general flow of the models at this level is as follows:

- At a start of the conflict, JEIM labor is deployed from the units in the *F-15 World* and the *F-16 World* blocks to the *FSL* block (*Labor Deployed* connections).
- Damaged engines from the FOLs (the *FOL* blocks are located inside the *F-15 World* and the *F-16 World* blocks) are sent through the transportation block (which contains an image of a truck) to simulate the transport delays.
- Engines leave the transportation block and enter the *FSL* block where they are serviced.
- FMC engines are sent back to the FOLs.
- At the end of the conflict, JEIM and Mod labor are removed from the *FSL* block and sent to the *Labor Reconstitution* block where they are sorted according to their place of origin (home unit).

RANDMR1614-3.3

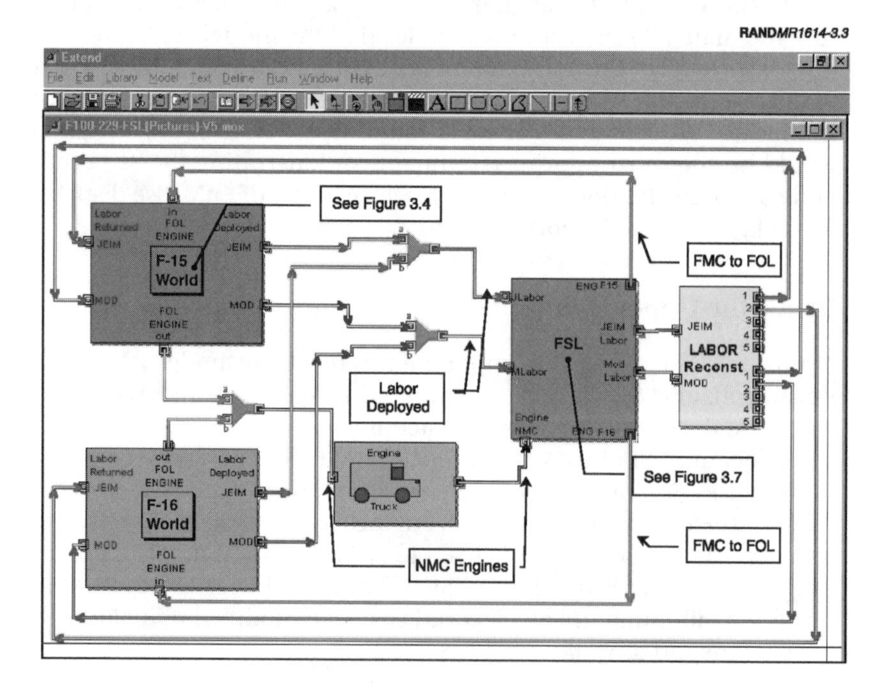

Figure 3.3—An FSL Block Model (top level)

- They are finally sent back to the *F-15 World* and the *F-16 World* blocks.

Each of the main blocks in the upper hierarchy constitutes a lower hierarchy of its own. For example, the *F-16* and *F-15 World* blocks shown in Figure 3.3 each contain several F-16 and F-15 bases, respectively; an *FOL* block where the forces are deployed; and other appropriate blocks such as *Transportation*. The *FSL* block contains the *JEIM* and *Test Cell* blocks. As indicated by the label "See Figure 3.4," which points toward the *F-15 World* block, we will next enter the lower hierarchy of the *F-15 World*.

Inside the Lower Hierarchies: An *F-15 World* Block

Figure 3.4 shows the contents of the *F-15 World* block, which was one of the blocks represented in Figure 3.3. The *F-15 World* block contains several F-15 bases (two shown), an *FOL* block where these

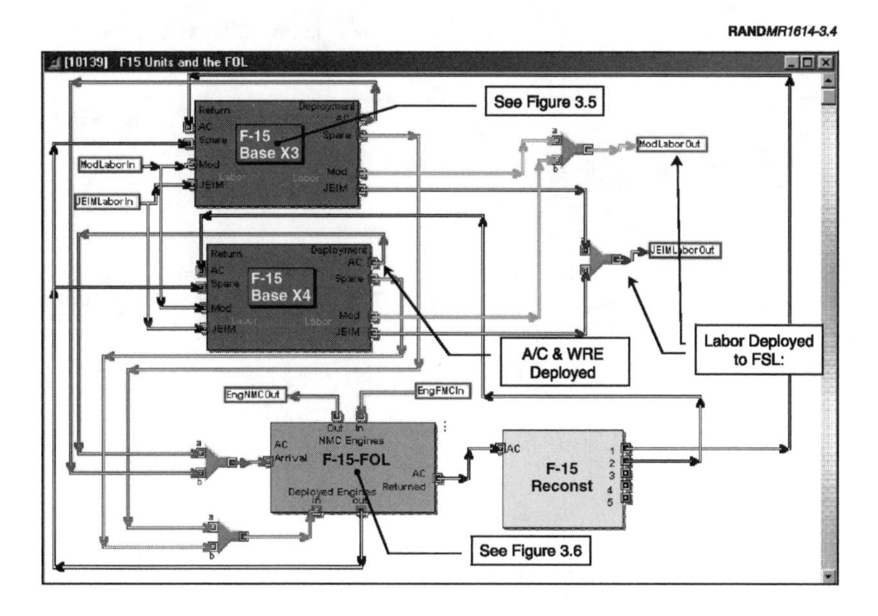

Figure 3.4—An F-15 World Block with Two Air Force Bases and a Single FOL

aircraft are deployed during a conflict, and a *Reconstitution* block. When a conflict starts, each base sends aircraft and war reserve engines (WREs) to the FOL, and JEIM and module labor (the latter two labeled as *JEIMLaborOut* and *ModLaborOut,* respectively) to the maintenance shop (e.g., an FSL).[6] After a conflict, aircraft, spares, and labor are reconstituted in the *F-15 Reconstitution* block, which returns aircraft, spares, and labor units to their original base.

We will now move inside one of the F-15 base blocks shown in Figure 3.4.[7] The *F-15 Base* blocks (Figure 3.5) and, later, the *FOL* block (Figure 3.6) will be used to show how the user-defined parameters are modified.

RANDMR1614-3.5

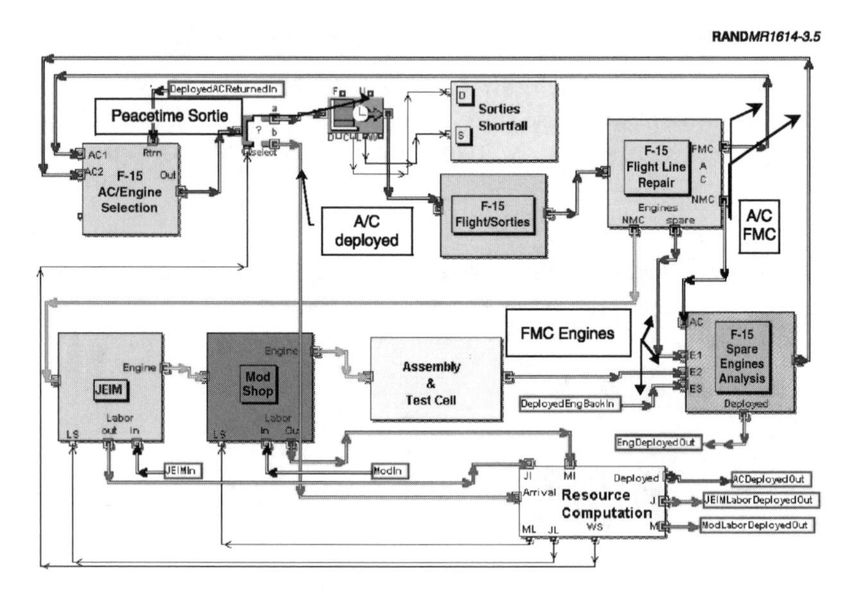

Figure 3.5—F-15 Base Block with a JEIM Shop

[6]Units deploy maintenance personnel to a deployed JEIM location or an FSL. In the CSL or Home Support scenario, no need exists for the movement of the personnel and therefore these pipelines are closed.

[7]Later in the chapter, we will explore the *FOL* block shown in the same figure.

An *F-15 Base* Block in Peacetime and Wartime Scenarios

As shown in Figure 3.5, the F-15 base contains several blocks that receive inputs from various parts of the model.[8] We will begin by briefly describing the function and interaction of the blocks and will then explain the inputs required to run the model in a peacetime scenario.

The *AC/Engine Selection* block tracks the number and types of aircraft (e.g., F-15, F-16) as well as the number of installed engines to be used in the simulation. The user enters these data into the model, while EnMasse automatically assigns the tail number, engine serial number, and engine cycle time. The aircraft and engines are combined to form FMC aircraft. They are sorted, based on the age of the engine, and are then queued for flight (exit the block).

After each sortie, aircraft are sent to the *Flightline Repair* block for inspection and maintenance. Aircraft that pass inspection are sent back to the pool of available aircraft. Some aircraft require minor repairs, which are performed on the flightline. Other engines are sent for scheduled or unscheduled maintenance. EnMasse pulls engines from the aircraft according to age and type of engine. For example, F100-229 engines have unscheduled engine removal (UER) rates and scheduled engine removal (SER) rates of 3.5 per 1,000 flying hours and 1.5 per 1,000 flying hours, respectively. Detached engines are tagged according to removal type and are sent to the *JEIM*. Aircraft tagged as NMC aircraft are sent to the *Spare Engines Analysis* block, where they are queued for the next available engine. These aircraft are either put back into service immediately, if serviceable spare engines are available, or they await the arrival of engines from the *Assembly and Test Cell* block.

The *JEIM* block processes NMC engines on a first-come, first-served (FCFS) basis.[9] Engines are first queued for parts and then for maintenance. Modular engines that have been processed by the JEIM are sent to the *Mod Shop* block. The *JEIM* block requires two inputs from

[8]The function and modification of each of these blocks is described in greater detail in Chapter Four.

[9]FCFS service discipline is modified in models with JEIM shop that serve both training missions and deployed forces (e.g., CSL) to give priority to the engaged forces.

the user: the initial number of labor units and the number of rails (i.e., the JEIM capacity).

The *Mod Shop* block separates modular engines for maintenance. There are five engine modules: fans, core, LPT, augmentor, and gearbox. In the current simulation, the gearbox is a two-level maintenance item that is sent to the depot. A portion of the other modules are also sent to the depot. The *Mod Shop* block requires several inputs from the user: an initial amount of labor, the capacity of the shop, and the spare level for each of the modules.

Engines leaving the *Mod Shop* block are sent to the *Assembly and Test Cell* block, where they are queued for assembly, test cell, and final inspection. The user is required to set the capacity of the test cell.

On leaving the *Assembly and Test Cell* block, the now-FMC engines are sent to the *Spare Engines Analysis* block, where they are pooled with the other spares (including the WRE) and available for installation on the aircraft. The FMC aircraft leave this block to join the pool of other aircraft and the whole cycle starts again. No user inputs are required in this block.

The *Sorties Shortfall* block keeps track of the daily demand and supply of aircraft and the daily number of missed sorties. This block is essential for measuring the performance of each scenario (see Figure A.1 for a sample output). The *Spare Engines Analysis* block keeps track of daily serviceable spares, the number of aircraft with holes, and the number and arrival time of serviceable engines.

An *FOL* Block

We will now examine an *FOL* block, which was part of the *F-15 World* block in Figure 3.4 and is shown in detail in Figure 3.6. The *FOL* block receives aircraft from other bases in the model according to the deployment schedule. As shown in the figure, the *FOL* block contains the *FOL Sortie Calculation* block, the *F-15 Flightline Repair* block, and the *Deployed F-15 Spare Analysis* block. Aircraft arriving at the FOL are queued and prioritized based on the health of their engines. The *FOL Sortie Calculation* block provides a user-defined schedule of sorties. Within this block we allow the user to set the sortie schedule explicitly (rather than using a simple daily rate) to allow for a more flexible sortie generation rate that reflects the reality that operators may demand different daily sorties. If this facility is used, daily sorties are determined by a tab-delimited external file with a single column of numbers representing the sortie requirement for each day. The *E-Time* Block, a user-defined block, signals the start of the engagement, at which time the arriving aircraft are pulled into the *Sortie Calculation* block.

As indicated previously in Figure 3.5, aircraft that pass the flightline inspection are sent back to the queue to await further sorties in the sortie calculation block. Otherwise, as in peacetime, the engines are detached and sent to a JEIM shop—in this case at an FSL, which is in another block because it is not located at the FOL. The *Deployed F-15 Spare Engines Analysis* block keeps the spare engines to be matched with the *aircraft* with holes. This block differs from its peacetime counterpart only in its ability to return the spares to the units at the end of the conflict, which is signaled by the *Reconstitution* block.

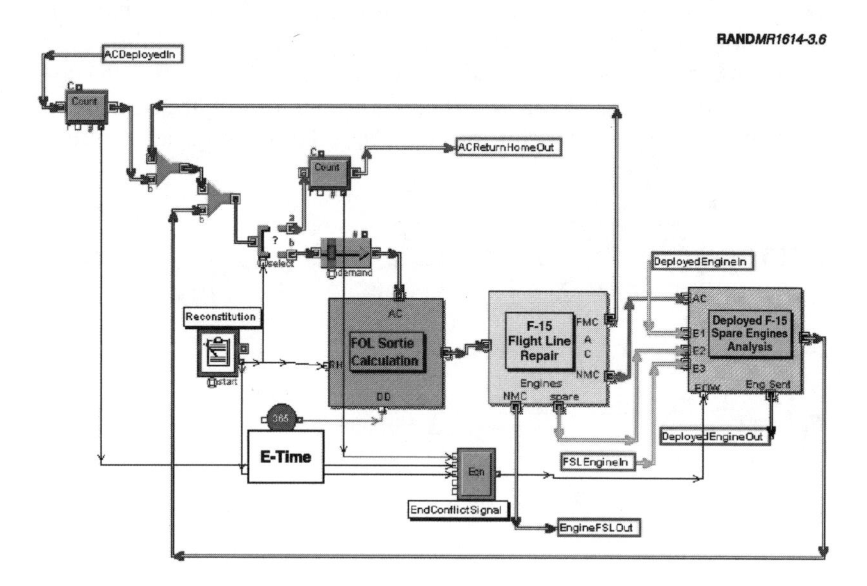

Figure 3.6—An F-15 FOL Block

A *JEIM Shop* Block

The *JEIM Shop* block can be in one or more of several parts of the model, depending on the scenario. The FSL model would have a JEIM shop at each home unit with another at the FSL; the CSL model would only have one JEIM shop; the deployed JEIM model would have a JEIM shop at each FOL and unit; and, finally, a home support scenario would have a JEIM shop at each unit home base supporting both engaged and nonengaged forces. Although the general structure of all JEIM shops is similar, some minor variations occur among the scenarios. Figure 3.7 depicts a JEIM shop in an FSL scenario. In Chapter Four, we will describe a JEIM shop using a CSL scenario as an illustration. Appendix B contains a complete list of JEIM shops.[10]

An *FSL* block is dormant during peacetime and becomes active only after receiving the deployment signal. At this point, labor units (*JEIM_Labor_In*) are deployed to the facility and start operating as soon as the first engine arrives (see Figure 3.7). A warm FSL would have some prepositioned labor, indicated by the *Organic Labor* label. The model combines both deployed and prepositioned labor into a single *Labor Pool.* At the end of the conflict, the *End_of_WarIn* signal is activated, the FSL shuts down, and the personnel are returned to their original units. During operation, disabled engines come in,

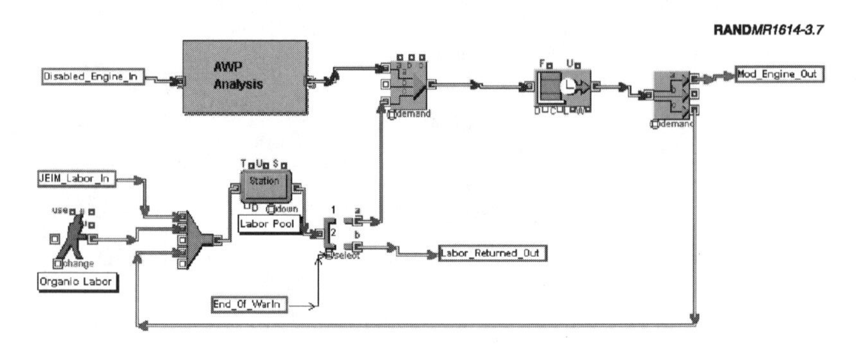

Figure 3.7—A JEIM Shop in a Forward Support Scenario

[10]See Figures B.1 through B.3.

potentially wait for parts (*AWP analysis*), and then are linked up with labor, if available. They have some repair done and then send modules to the collocated module shop and release the original labor to work on the next engine (follow the icons across the top of Figure 3.7).

The deployed JEIM is similar in structure to the FSL except for its location (collocated with the engaged forces) and the added delay in the test cell operation. Home Support and CSL scenarios involve both engaged and nonengaged engines, and therefore JEIM shops in these scenarios require additional features: the ability to switch from a peacetime workday to a wartime workday and the ability to give priority to the engaged forces.

RUNNING THE MODEL USING THE DEFAULT SETTINGS

The run time depends on the number of entities in the system. For a relatively large number of aircraft and engines (about 620 aircraft and 1,000 engines), the model may take up to three minutes to simulate two years of activity.

At the end of each run, every block generates an output (a single number or any array of data, depending on the type of block) that can then be read by the user. Extend allows these outputs to be captured by an external file using its report generator capacity.[11] The data from several runs can be captured without any manual intervention.

The next chapter is devoted to a more detailed description and potential modifications of the EnMasse modules. We will show how the flexibility of the model allows it to be adapted for further analysis of engine maintenance.

---

[11]This is done by highlighting the desired block and then selecting *Add Selected to the Report* under the *Run* pull-down menu. For more information, see the Extend user manual.

Chapter Four

EnMasse LIBRARY

This chapter provides a detailed presentation of the contents of the EnMasse library. The library holds all the objects needed to build a maintenance model for jet engines, whether modular or nonmodular. We examine each of the blocks that make up the engine maintenance process, as well as the blocks used to simulate such operations as the transport of engines and the joining of serviceable spares with aircraft. In general, each block either pulls data from or pushes data toward other blocks, using an input or output pipeline, as required. Some blocks need initial input (drawn from user input or default settings). Each block generates an information output (e.g., length of queue) that can be captured by an external file, depending on the analysis being done.

The remainder of this chapter describes the purpose of each block in the library, its usage, and the flow of entities through it (input and output pipelines). Information is also provided on how the user can modify the blocks, if desired, to enhance the model's functionality. The idea behind this chapter is to provide enough information about the capability of each module to allow the user to build a new set of models capable of conducting different assessments of engine maintenance than those shown here.

AIRCRAFT/ENGINE SELECTION BLOCK

This block is illustrated in Figure 4.1. The *Aircraft/Engine Selection* block tracks the number and types of aircraft as well as the number of installed engines to be used in the simulation. This block also serves as a holding cell for fully mission capable (FMC) aircraft coming from other parts of the model (e.g., aircraft returning at night, engaged aircraft coming home). *Aircraft/Engine Selection* blocks come in two types: a single-engine and a two-engine model. We describe only the two-engine block because the single-engine is very similar.

The main processes of this block can be seen as we move from left to right in Figure 4.1. The initial number of *aircraft* and engines (not including spares) are defined by the user, while EnMasse assigns the tail number, engine serial number, and initial age (other attributes may be added in the *Set Attributes* block). Aircraft are prioritized according to the age of their engines for sortie service and then queued for processing by the *Sortie Generation* block (see below). The *Aircraft/Engine Selection* block, as well as being the starting point for aircraft, is also a gathering point for aircraft that have been put back into the fleet after being rated not mission capable (NMC) (*FMC_AC_In*) and, when representing a home base, aircraft continuing with training sorties (*AC_back_to_Action_In*) and aircraft returning from deployment (*Deployed_AC_Return_In*).

Input: The user enters the initial number of *aircraft* (*Initial Aircraft* stock) and the initial number of FMC engines (*Initial Engines* stock). The model sets the default values for tail numbers, engine serial numbers, and age, any of which may be modified by the user through the *Set Attribute* block. Figure 4.2 illustrates such a modification. The data for aircraft and engine attributes are read from an

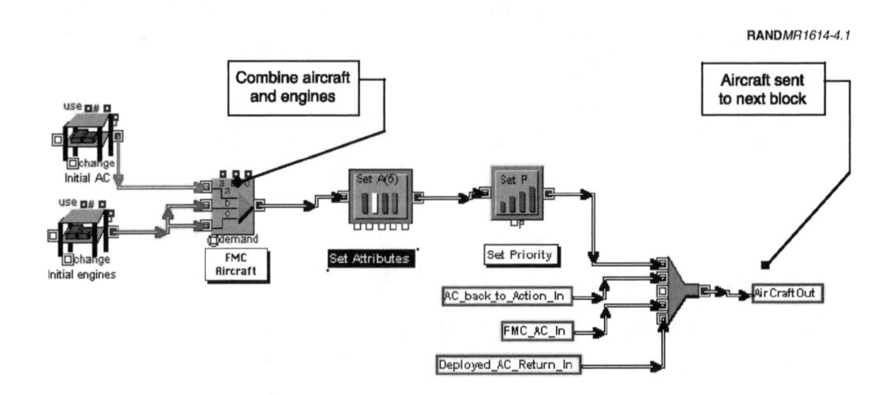

Figure 4.1—Aircraft/Engine Selection Block

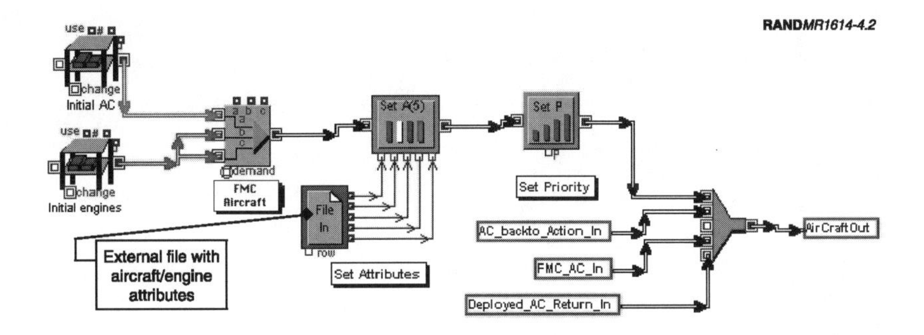

Figure 4.2—Aircraft/Engine Selection Block with User-Modified Attributes

external file (such an Excel spreadsheet or an ASCII file). Each pipeline from the *File Input* block corresponds to a column in an external file (see Figure 4.3).[1]

This block receives aircraft data from three sources: aircraft continuing with training sorties (*AC_back_to_Action_In*), aircraft that have been put back into the fleet after being rated NMC (*FMC_AC_In*), and returning deployed aircraft (*Deployed_AC_Return_In*).

Output: This block sends FMC aircraft to other blocks. It does not, however, generate any output for purposes of analysis.

FOL SORTIE CALCULATION BLOCK

The *Sortie Calculation* block, shown in Figure 4.4, takes as input FMC aircraft (based on available airframes and engines) and attempts to have those aircraft fly a variable set of daily sorties during a conflict. By default, the model is set to read an external file with the daily sortie schedule used to compute the aircraft. The *Sortie Calculation* block also has its own *Sorties Shortfall* block that keeps track of the performance of the FOL. The *Sortie Calculation* block also adjusts such flight attributes as engine clock and provides a priority number to each aircraft based on its age. Aircraft leaving this block are sent to the *Flightline Inspection* block. At the end of the conflict, the

[1]Details of reading in files from Excel can be found in Extend documentation.

RANDMR1614-4.3

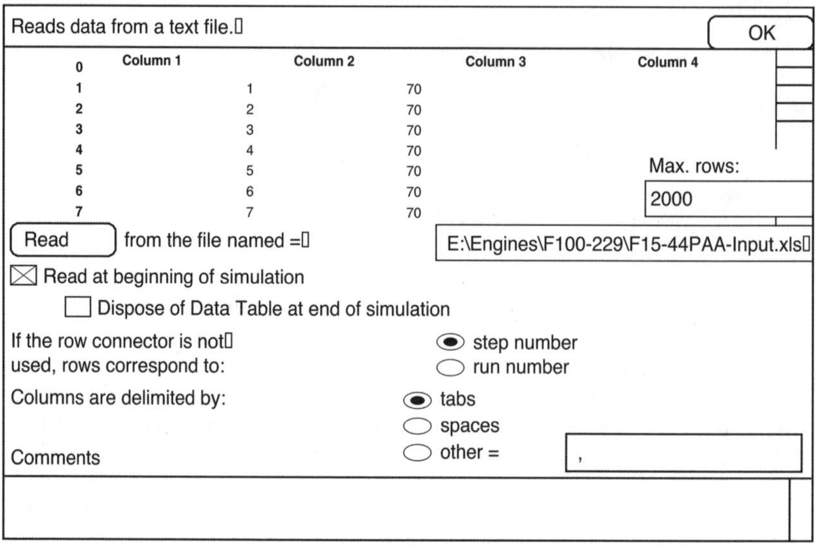

Reads data from a text file.

OK

0	Column 1	Column 2	Column 3	Column 4
1	1	70		
2	2	70		
3	3	70		
4	4	70		
5	5	70		
6	6	70		
7	7	70		

Max. rows: 2000

Read from the file named = E:\Engines\F100-229\F15-44PAA-Input.xls

☒ Read at beginning of simulation

☐ Dispose of Data Table at end of simulation

If the row connector is not used, rows correspond to: (●) step number ( ) run number

Columns are delimited by: (●) tabs ( ) spaces ( ) other = ,

Comments

Figure 4.3—External File Input

ReturnHomeSignalIn is switched on and the aircraft are pulled back to their original bases.

If the user prefers to avoid use of an external file, a simple modification to the *FOL Sortie Calculation* block will allow internal sortie demand generation, as shown in Figure 4.5. In the figure, this modification has been made with the addition of two parameters: *MDSIn* (an output from another block indicating the number of *aircraft* deployed) and *Stress Factor* (a new user-defined input to indicate the number of days at surge rate). An equation (*Eqn* block) computes the number of *aircraft* needed based on the clock time, number of *aircraft* deployed, and number of days of surge.

Input: An external, tab-delimited file, such as an Excel file, is required. The *File In* block can read up to five columns of data, although for the current model we need only the first two columns. The user needs to create a data file with the first column indicating

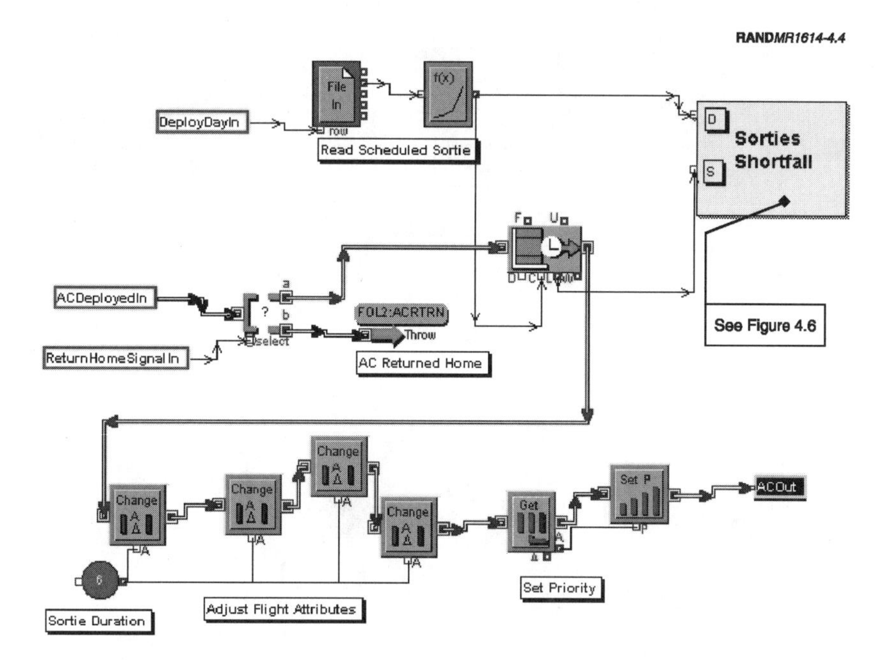

Figure 4.4—FOL Sortie Calculation Block

the days of conflict and the second containing the number of corresponding required sorties.

The sortie duration is a constant set based on two sorties per day (e.g., six hours indicates an average of two three-hour sorties per aircraft). This set can be modified in the *Sortie Duration* block. No other user input is required for this block. However, the modified block (Figure 4.5) requires an input from the user indicating the number days at surge rate.

The *FOL Sortie Calculation* block has three input connections (i.e., requiring data from other blocks): a binary signal (*ReturnHomeSignalIn*), which makes decisions about the state of the system (i.e., war or peace); deployed aircraft throughput (*ACDeployedIn*); and a daily signal (*DeployDayIn*), which counts the days of deployment. In the modified block, an additional signal (*MDSI*) keeps track of the total number of aircraft deployed.

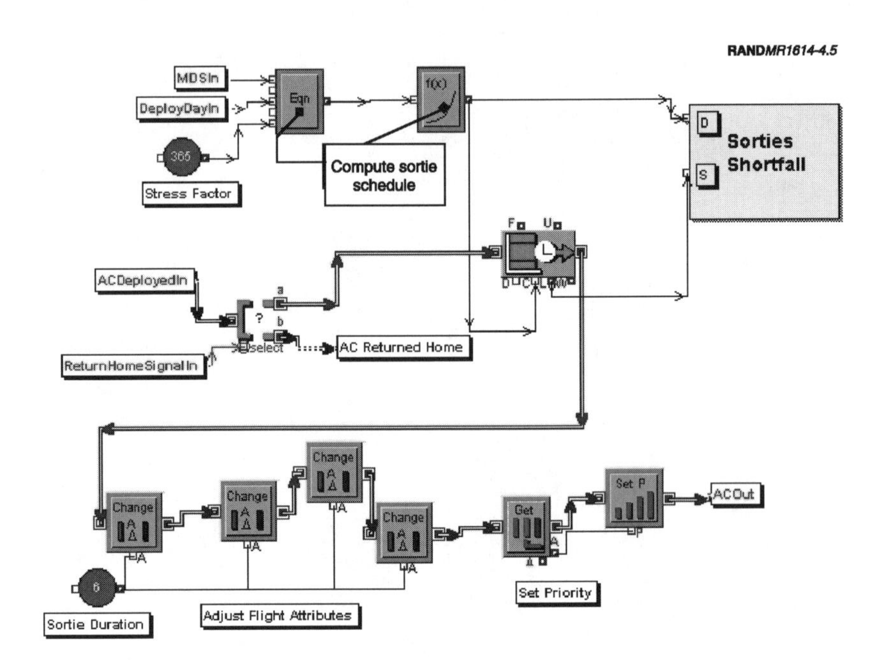

Figure 4.5—Modified FOL Sortie Calculation Block

Output: No data for analysis purposes are produced at this level. (The lower level, the *Sortie Shortfall* block, is discussed below). This block has one output connection (i.e., producing data for other blocks) that pushes aircraft entities to the next block.

SORTIE SHORTFALL BLOCK

The *Sortie Shortfall* block tracks daily sortie losses by looking at the required number of sorties based on the utilization rate in peacetime or the scheduled mission in wartime, the aircraft availability, and the actual number of sorties flown. The model keeps track of daily requirements and computes both the average and daily demand as well as supply for the system. The graph in Figure 4.6 depicts the average result of several simulation runs for an F-15 FOL with the initiation of the conflict on day 365 and termination on day 560 (see Figure A.1 for an F-16 FOL result).

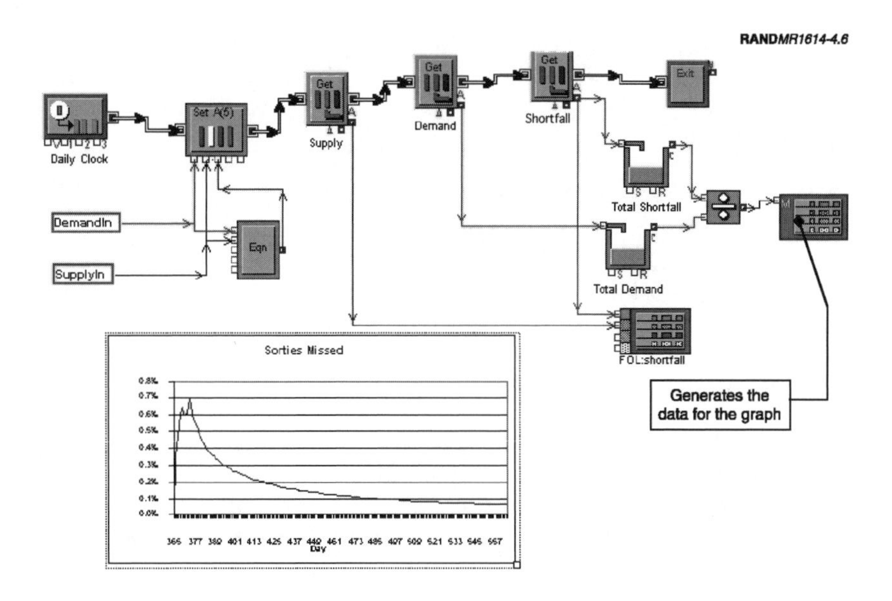

Figure 4.6—Sortie Shortfall Block

Input: No user input is required. The *DemandIn* and *SupplyIn* blocks receive inputs from the *Sortie Calculation* block (Figure 4.4).

Output: This block generates sortie information needed to analyze the performance of the system. The user may observe the system during each run and receive a report on the daily and average sortie performance. Two charts are kept at each run. One tracks the daily number of sorties lost and the other the percentage of sorties lost (shown in figure). The model also tallies the total number of *aircraft* needed and the system shortfall. This block does not have any output connection to another block.

FLIGHTLINE BLOCK

The *Flightline* block simulates the flightline inspection and determines the need for scheduled and unscheduled maintenance. Each engine, according to its age or some probability distribution, is tagged for flightline maintenance, scheduled/unscheduled maintenance,[2] or no maintenance (see Figure 4.7). Flightline inspection and repair are done without pulling the engine. After flightline maintenance is complete, the cycle clock usage (not shown) is reset and the aircraft is returned to the FMC pool (*AircraftFMCOut*).

Scheduled maintenance and unscheduled maintenance both necessitate engine removal and are performed by the JEIM shop. Engines that require JEIM are tagged accordingly and removed from the aircraft. When this kind of maintenance is performed, aircraft are designated as NMC and must await working engines (either spares or repaired engines). Aircraft not in need of engine maintenance are returned to the pool of FMC aircraft.

Input: This block requires three sets of user-defined data: flightline inspection rate, scheduled removal rate, and unscheduled removal rate. The user may modify the flightline inspection threshold by editing parameters in the equation block (*Eqn*) to change the inspection interval and may change the removal rates by editing the

RANDMR1614-4.7

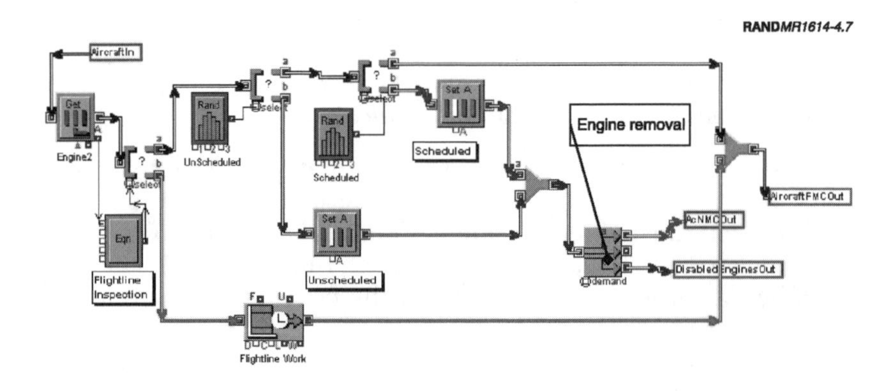

Figure 4.7—F-16 Flightline Inspection and Repair Block

[2]Initially we had intended to separate scheduled and unscheduled maintenance and had written the model to do so. However, data analysis indicated that we could not accurately determine the times required for each type of maintenance (the coding for any given maintenance action is often that for unscheduled maintenance, even when most of the work is scheduled). Also, the set of scheduled maintenance activities and their different time cycles was quite complex and would not have affected the study we were doing; this approximation was sufficient.

parameters in the random number blocks (*Rand*).[3] Additional attributes could be added[4] to the engine if it is subjected to an unscheduled removal to distinguish among the various types of unscheduled removals (e.g., minor FOD, major electrical problem). These attributes can then be used in the modified JEIM model to adjust length of time to fix, etc., although this would require much deeper data analysis than we performed for our study. There is one input connection (*AircraftIn*) for receiving aircraft from other blocks.

Output: Throughout the simulation, this block performs a daily tally of the number of engines removed, the number of flightline inspections completed, and the number of aircraft with holes. These are available for external analysis and were logged for some of our analyses. There are three output connections to other blocks: FMC aircraft, NMC aircraft (sent to the *Spares Analysis* block, Figure 4.16), and engines requiring JEIM.

JEIM SHOP BLOCK

The *JEIM Shop* block can perform both scheduled and unscheduled maintenance. As shown in the top portion of Figure 4.8, engines arrive (potentially from engaged forces *[FOL_Engine_In]* and nonengaged forces *[Disabled_Engine_In]*) and go through the *Awaiting Parts* (AWP) *Analysis* block,[5] where a decision is made whether to delay the engine because of lack of parts or send it directly to maintenance.[6] Engines that leave the *AWP Analysis* block are queued to be assigned to a labor unit (*Labor Pool*).

Figure 4.8 depicts a JEIM shop in a CSL where both engaged and nonengaged forces are maintained. At the start of the conflict, the

---

[3]In this model, an engine is designated for removal (either scheduled or unscheduled) by generating a Bernoulli (0 or 1) random variable with probability of 1 (removal) equal to the removal rate per flying hour multiplied by the number of flying hours logged in the last sortie. Other probability distributions could be used by modifying or replacing the random number blocks.

[4]Extend allows attributes to be added or removed to objects as they move through the simulation.

[5]The ENMCS block is labeled *AWP* block in the simulation model.

[6]The *AWP Analysis* block is described in full below.

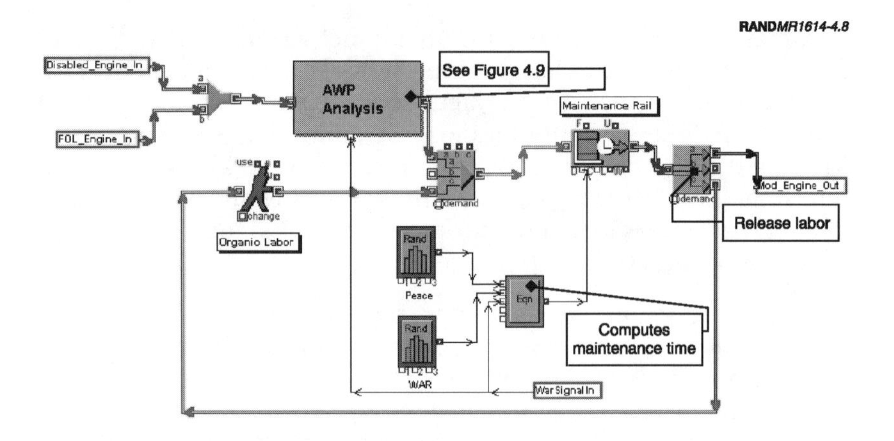

Figure 4.8—JEIM Shop Block in a CSL

War Signal is activated and the shop shifts to a 24-hour-a-day, seven-days-a-week work operation. Otherwise, work is done on a two eight-hour shift, five-days-a-week schedule. The in-work process time at the JEIM depends on the type of engine and, by default, is calculated by a probability distribution based on historical data. As will be explained in the JEIM Modification section of this chapter, the in-work time can be modified to reflect the experience level of the labor mix or scheduled/unscheduled engine removals.

Input: The user must set the number of labor units and the maintenance duration. The initial (prepositioned) amount of labor is assigned in the *Organic Labor* block). The in-work distribution is computed in the random block and is based on a peacetime or wartime work schedule. This block receives engines and labor from other blocks as well as a signal to switch to a wartime schedule.

Output: The queues before each activity as well as the utilization of the rails (*Maintenance Rail*) and labor (*Organic Labor*) may be logged either in the model and displayed or output externally for more complex analysis (we generally used these as diagnostics for which the former was sufficient). The model keeps track of the number of engines entering and leaving the JEIM. The output pipe depends on the type of JEIM: an FSL and deployed JEIM will have two output pipes, which return labor and engines leaving the shop; a JEIM in the CSL or Home Support mode has one output: engines. A JEIM in a unit will deploy engines with the engaged forces and may, if the scenario has an FSL, also deploy labor to the FSL.

AWP BLOCK

The *AWP* block (Figure 4.9) is contained within the *JEIM Shop* block. The *AWP* block adds a delay to the process to simulate the time spent awaiting parts for engine repair. Engines enter the block and, if the required parts are not available, are routed to a holding cell as ENMCS. In our study, the ENMCS calculation was based on historical data, which was used to construct an empirical distribution of waiting times (see Figure A.3). An engine that is no longer ENMCS exits the block to be worked on by the labor units.

Input: The parameters in the *ENMC-S Distribution* and *Delay Distribution* can be edited by the user to modify the AWP probability distribution. This block receives engine entities from other blocks.

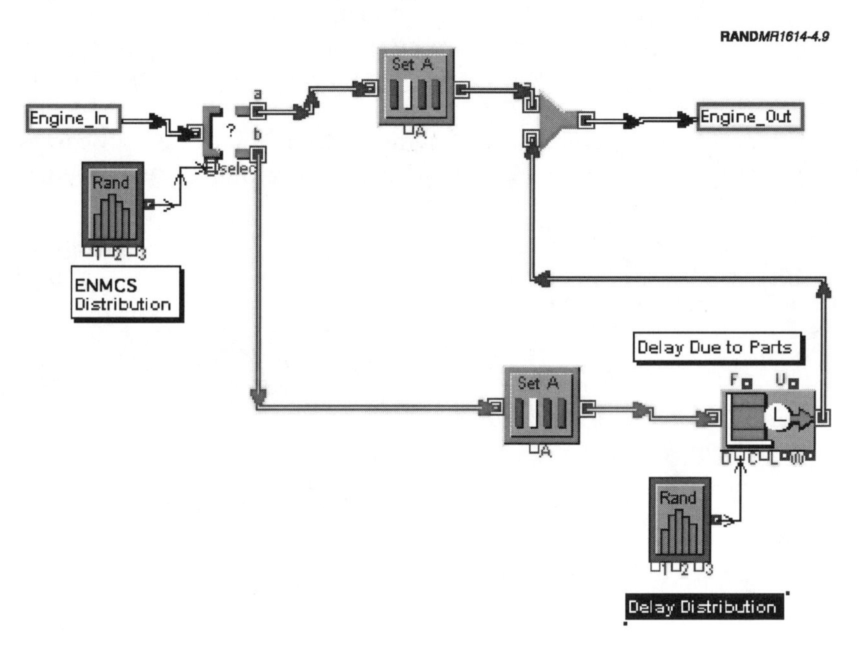

Figure 4.9—AWP Block

Output: As with the previous block, the user may observe the ENMCS queue and the wait times for engines. This block pushes engine entities to other blocks.

MODIFIED JEIM BLOCK

As indicated earlier, the *JEIM Shop* block can easily be modified to accommodate the experience level of the labor pool or the status of the engines (e.g., if removed because of scheduled or unscheduled maintenance). Figure 4.10 depicts a modified JEIM shop for an F-15 base. The labor pool consists of personnel at the base and returning labor from an engagement. The unit JEIM deploys personnel to the field of operations via the *Labor_Deployed_Out* block. An engagement is signaled by the *LaborWarSignalIn* block.

In the modified JEIM, the flow of the process is as follows. The base labor is assigned a new attribute level known as an experience level in the *Experience Mix* block (the returning labor has already been assigned an experience attribute by this or other units). The labor pool therefore contains a mix of personnel with various levels of work experience (e.g., E3, E5). As with all other types of JEIM blocks, labor and NMC engines (after leaving the *AWP* block) are processed on a rail. In the modified block, however, the maintenance time is a

RANDMR1614-4.10

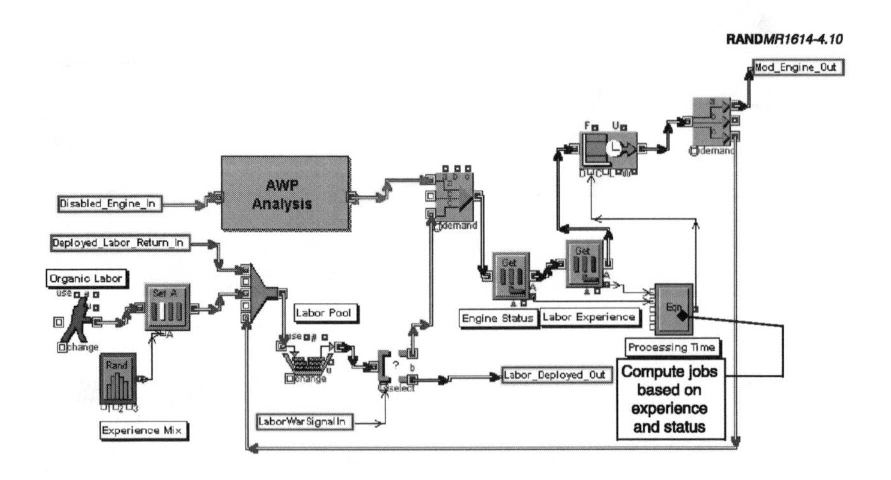

Figure 4.10—Modified JEIM Block

function of status of the engine (i.e., scheduled or unscheduled maintenance) and the experience of the team dealing with the maintenance (indicated by *Engine Status* and *Labor Experience,* respectively). The equation block (*Processing Time*) combines these two attributes to create a delay in the process. The system keeps track of the number of engines arriving and the time of their arrival. At the end of this process, the engines are pushed to the next block and the labor is released and returned to the labor pool.

Input: The user may define the experience mix and delay time by modifying the *Experience Mix* block and the *Processing Time* block, respectively. This block receives engine entities and labor entities from other blocks.

Output: The output for this block is similar to that of the JEIM mentioned earlier. The user may observe the queues before each activity as well as the utilization of the rails and labor. The model keeps track of the number of engines entering and leaving the JEIM. The output pipes for this type of JEIM are deployed labor and engines (or modules, depending on the type of engine).

MODULE SHOP BLOCK

For such modular engines as the Pratt & Whitney F100 series, a *Module Shop* block must be added to the model. The engines leaving the *JEIM Shop* block enter this block and are immediately sorted into three categories: modules sent to the depot, modules awaiting maintenance because of parts, and modules ready for maintenance.

We have incorporated the parts delay for modules in the *AWP Analysis* block of the JEIM shop. (However, if the user wishes to represent the AWP explicitly for modules in the module shop, an *AWP Analysis* block can be added immediately after the *Module_Engine_In* pipeline.) Modules that require depot maintenance are pushed through the *Depot Maintenance* delay process (i.e., depot maintenance is not explicitly represented in the current model). Other modules are assigned to labor when available and enter the *Mod Repair* block (or *Deployed Mod Repair* for module shops that can receive labor from other parts of the model). Figure 4.11 depicts a

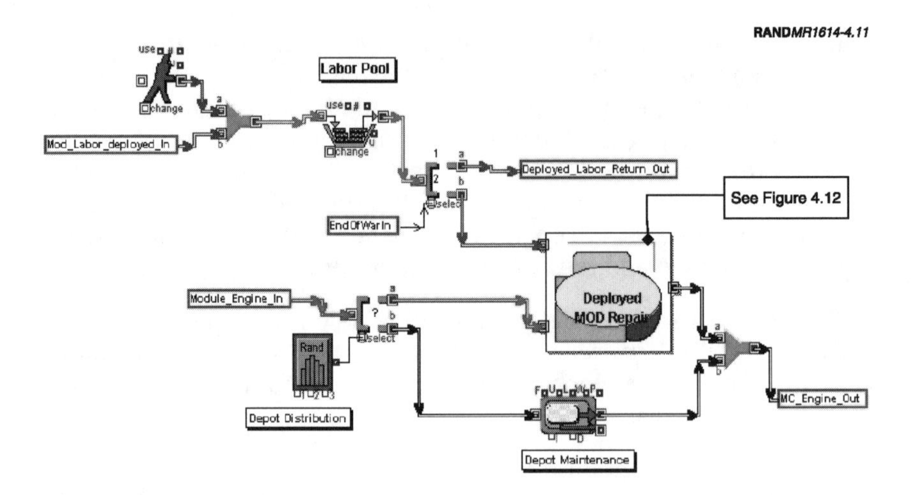

Figure 4.11—Module Shop Block

Mod Shop block at a deployed location (FSL or FOL) where prepositioned labor is pooled with the deployed labor to form a *Labor Pool.*

In the module shop, engines are separated into five modules to be repaired at five parallel work centers (see Figure 4.12). By default, the gearbox is sent directly to a depot; therefore, its block is more of a placeholder. The other blocks pull in labor and modules for repair, with each having its own spare capacity and repair capability.

Figure 4.13 illustrates the shop capabilities for the core module repair shop. Arriving engine cores are divided into healthy parts that are released immediately and those that require attention. The latter group is queued for repair. The repaired cores are pushed to the next stage of maintenance. The other shops have similar structure.

Input: The *Module Shop* requires several inputs from the user: the initial number of personnel (prepositioned labor), the capacity of each *Mod Shop*, and the spare level at each *Mod Shop*. The user may also want to modify the parameters in the *Depot Distribution* block, which determines the percentage of modules going to a depot, the repair time at each of the *Mod Shops*, and the distribution of healthy modules entering the depot shops.

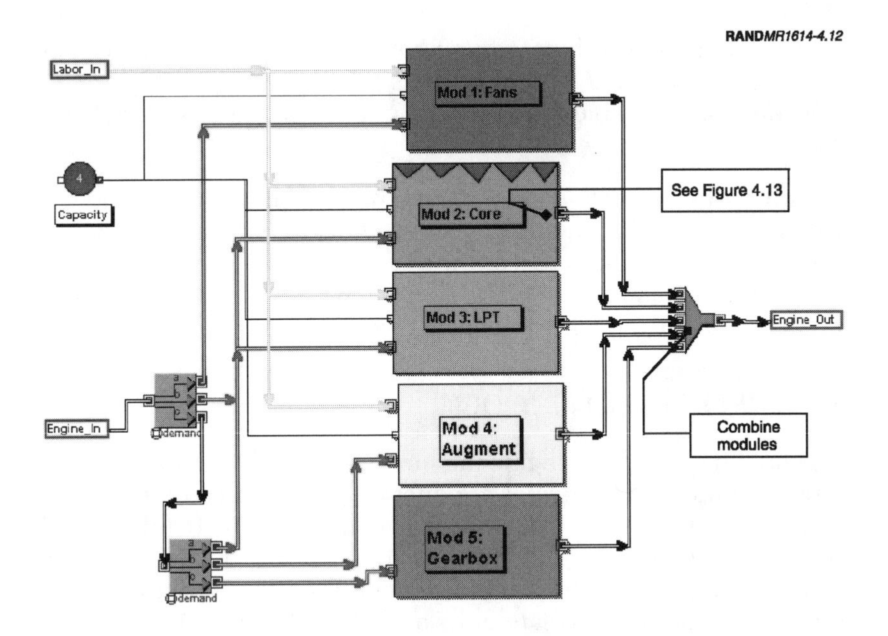

Figure 4.12—Module Repair Block

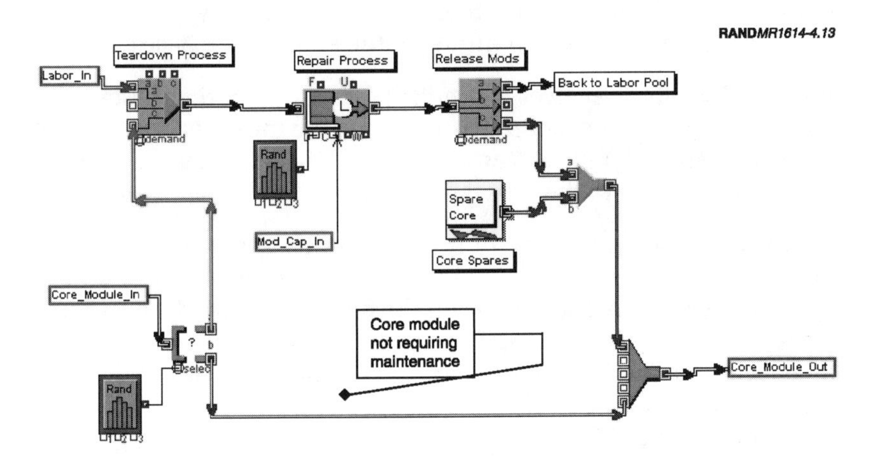

Figure 4.13—Core Module Repair Shop Block

Two pipelines enter the *Module Shop* block: labor enters through *Mod_Labor_Deployed_In* and engines through *Module_Engine_In*. The *Module Repair* block receives labor and modules from the *Module Shop* block.

Output: The *Mod Shop* block keeps track of queue length and size for the labor and the modules. The *Mod Repair* block keeps track of the shop's utilization as well as the queue length and size for both the labor and the modules. The *Module Shop* has two output pipelines to the rest of the model: engine and labor unit output.

ASSEMBLY/TEST CELL BLOCK

The final block for the engine maintenance is the *Assembly/Test Cell* block. In this block, engines are queued for final assembly, the test cell, and final inspection. As shown in Figure 4.14, this block includes several delays:[7] for the final assembly, the "hush house" (the test cell stand), the final inspection, and the rerouting of the engines to their appropriate units or FOLs.[8] The flow is as follows: modules are assembled (about one to two days) and then queued for the hush house. The majority of the engines require about one day of delay, but a small portion may take up to two days. Finally, the engines are inspected and returned to the appropriate bases.

Input: This block receives engine entities from another block, usually the *JEIM* or *Mod Shop* block. The user must set the test cell capacity by modifying the *Test Cell* block. The user may also modify the delays for each of the processes.

Output: The main analytic outputs from this block are the queues before the test cell and the utilization of the test cell (we simply recovered summary statistics from the block). The model also keeps track of the number of engines served and the utilization of the assembly and final inspection blocks. The output pipeline depends on the number of bases or FOLs. In Figure 4.14, we have depicted a scenario with four FOLs.

---

[7]These delays were based on data analysis or on interviews with JEIM personnel.

[8]Figure 4.14 depicts a scenario with two F-16 and two F-15 FOLs.

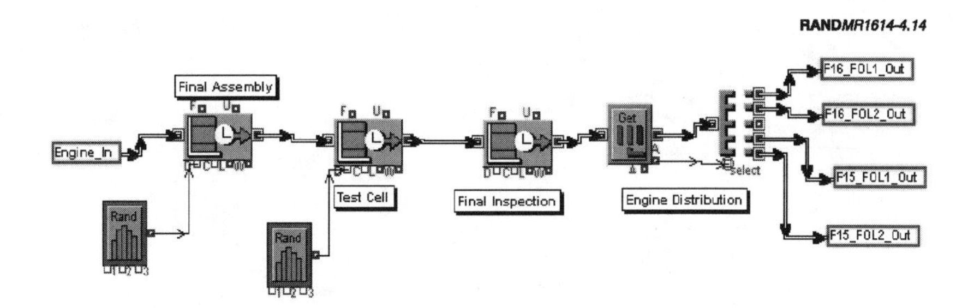

Figure 4.14—Assembly and Test Cell Block

SPARES ANALYSIS BLOCK

The *Spares Analysis* block, housed in an FOL or base block, receives FMC engines from the *Assembly/Test Cell* block. NMC aircraft from the *Flightline* block are also sent to this block where they are stripped of some of their attributes and queued for serviceable engines.

Engines come to the *Spares Analysis* block from various sources depending on the Mission Design Series (MDS) and the type of spares analysis (i.e., FOL or unit). Figure 4.15 depicts a spares analysis block for an F-15 FOL (see Figure A.6 for a sample output). Serviceable engines arrive at the block from the *Assembly/Test Cell* block (*Repaired_Engine_In*) while cannibalized engines come from aircraft with only one inoperable engine (*Canned_Engine_In*) and from deployed engines (*Spares_Engine_In*). Aircraft with holes are joined with the appropriate number of engines (i.e., depending on the MDS) to form FMC aircraft and then sent to the next block. The end-of-war signal (*EOW_In*) indicates when the deployed engines are to be returned to their original bases.

There are two different types of *Spares Analysis* blocks, depending on whether a deployed location or a base is used. The deployed location is depicted in Figure 4.16, where, after the conflict, WREs are returned to the home bases. Figure 4.17 depicts a block for an F-15 base where WREs are instead deployed to various FOLs. In this block, a *Spare Engine* block has replaced the "SpareEngineIn" input to account for the possibility that units might have their own reserve serviceable spares. In addition, because it is the start of war that signals the deployment of the WREs for units, the *EOW_In* and

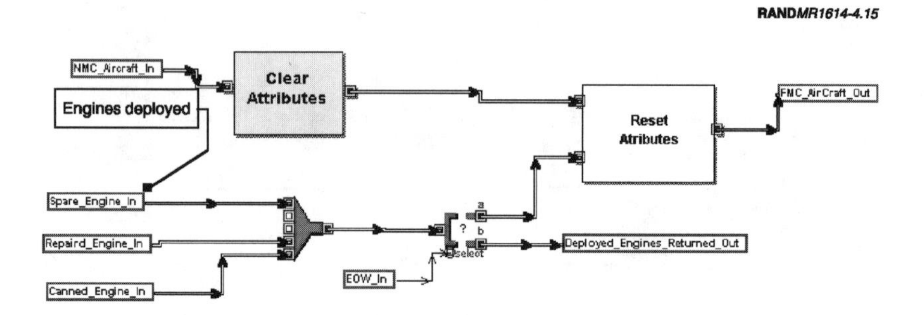

Figure 4.15—Spares Analysis Block at an FOL

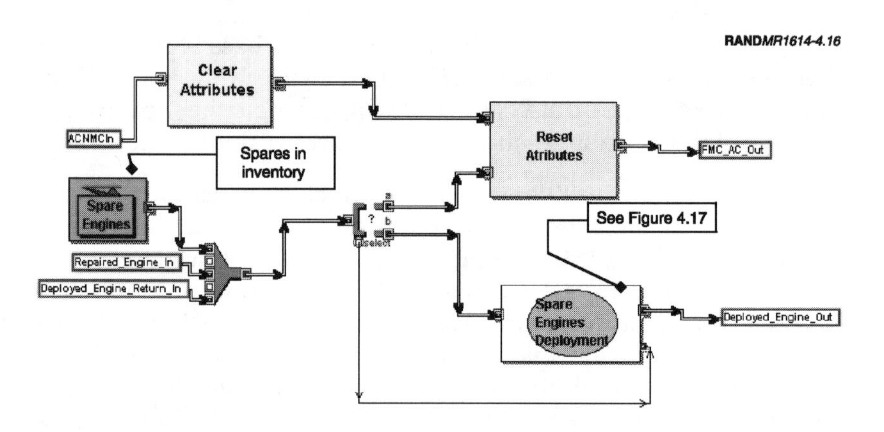

Figure 4.16—Spares Analysis Block at an F-15 Base

Deployed_Engine_Return_Out blocks have been replaced with a *Spare Engines Deployment* block, which keeps track of the number of engines deployed and the date they are deployed (see Figure 4.17). Finally, because the units receive engines from deployed locations ("Engines Deployed" note), this block has an additional pipeline to receive the returning engines from the FOL.

Input: For Spares Analysis at bases (Figure 4.16), the user must enter the initial number of spares, the number of WREs (i.e., engines to be deployed), and the day of deployment (*Spare Engines Deployment* block). The *Spare Engines Deployment* block requires two inputs

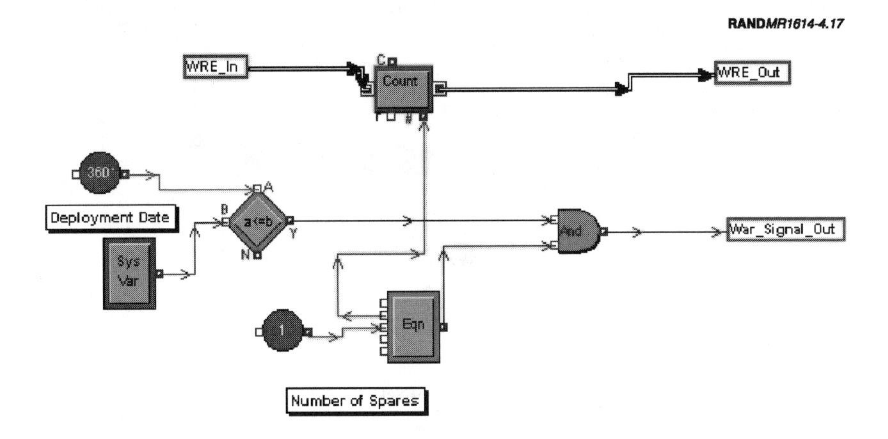

Figure 4.17—Spare Engines Deployment Block

from the user: the number engines to deploy (WRE) and the day they should start deploying. The *Spares Analysis* block for the FOL does not require any specific user inputs.

Output: Several outputs are produced by both types of blocks. There are two output pipelines, independent of the type of the block: FMC aircraft and the engine entity output. These return engines for the FOL block and deploy engines for the unit block. Additionally, the model keeps track of daily spare levels, the arrival time of aircraft with holes and serviceable engines, the queue length and time for aircraft with holes, and the length of time an engine is tagged as NMC.

TRANSPORTATION BLOCK

One of the main features of our model is the explicit representation of transportation, both for intertheater and intratheater. This block consists of a pool of vehicles (ground or air) and a one-way transportation time. Figure 4.18 illustrates the transportation route for round-trip travel. Engines enter the block, where they are joined with vehicles (from the "Vehicle Free" stock, each of which has a different capacity). At the end of the travel, vehicles and engines are released.

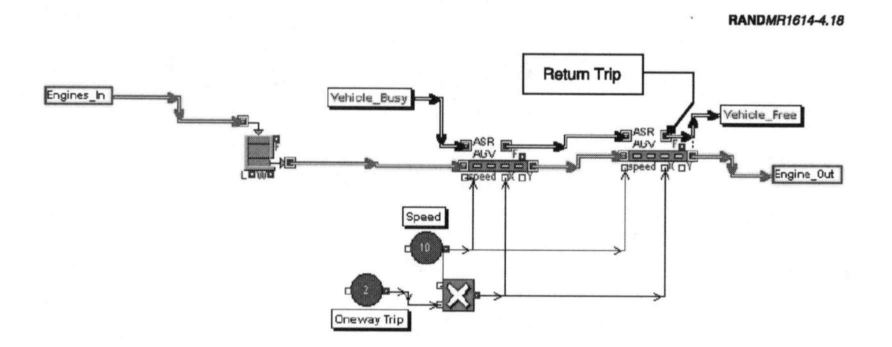

Figure 4.18—Transportation Block

Input: The user must enter a value for the expected one-way travel time (*One-way Trip* block), as well as the number of vehicles and their capacity. If the user requires a range of time or a location-dependent time, the *One-Way Trip* block must be replaced by either a *Random* block or a block that can respond to location attributes. Engines and vehicles are pulled from other blocks.

Output: The model measures vehicle utilization as well as the queue length and time for the engines. Engines and vehicles are pushed to other blocks.

RESOURCE COMPUTATIONS BLOCK

The *Resource Computations* block keeps track of the deployment of labor and aircraft. The *Deployment Day* block, a user input, indicates the start of the deployment for labor and aircraft (see Figure 4.19).

The *Deployment Delay (Labor)* block is provided to capture the absence that some engine personnel not deployed with the fighter aircraft. The *Deployment Calculation* blocks produce a binary signal indicating whether the desired number of aircraft or labor has been satisfied (see Figure 4.20). The deployment date (including any delay) is compared with the current date (system clock) and on the desired day the *Deployment_Signal_Out* block is activated, telling the receiving block to release aircraft or support personnel. The aircraft and labor are passed through this block, generically identified as

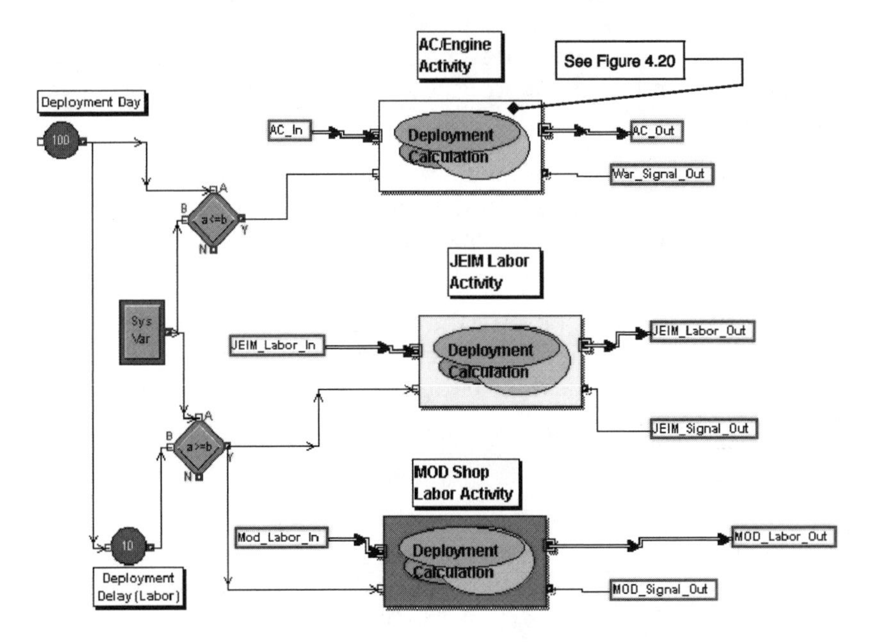

Figure 4.19—Resource Computations Block

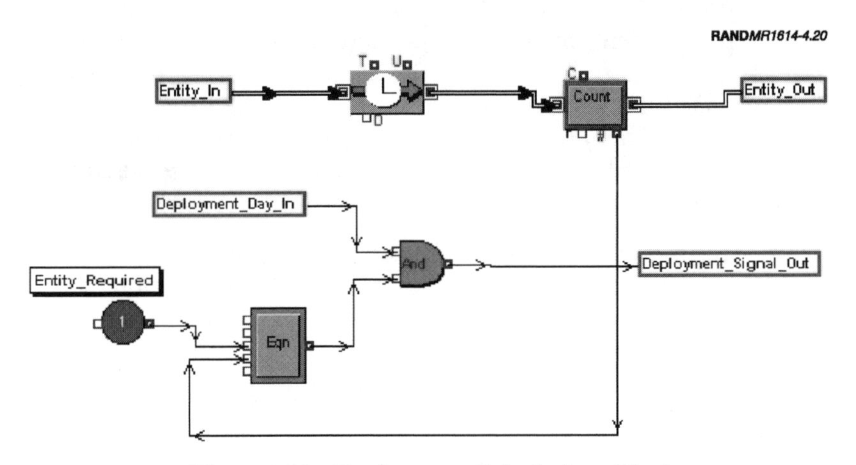

Figure 4.20—Deployment Calculations Block

Entity_In, and their count is compared to the desired number, *Entity_Required*. As soon as the desired number is reached, the *Deployment_Signal_Out* is deactivated. This, of course, does not indicate the end of the conflict but rather the completion of the deployment.

Input: The user must input the day of deployment, possible delay in labor deployment, and number of aircraft or labor required (*Deployment Calculations* block). Labor and aircraft are pulled from other blocks.

Output: The user may observe the number of entities going through the *Deployment Calculations* block, but no analysis tool is in this block. Labor and aircraft signals and labor and aircraft entity are the outputs for this block.

FOL BLOCK

The final block in the library is the *FOL* block, which receives aircraft from other bases and holds them until the end of an operation (Figure 4.21). This block is similar to those found in units in terms of structure but lacks any organic resource. The aircraft are deployed to this block from other units, based on the numbers from the *Resource Computation* block. They are queued until the *E-Time* signals the operation, at which time they are pushed to the *FOL Sortie Calculation* block, *Flightline* block, and the *Spare Engines Analysis* block. The *Reconstitution* block signals the return of the *aircraft* to their original bases. The "Return Complete" equation does not signal the end-of-war until all the aircraft are returned. At that time, it activates the return signal for the spares and personnel (*EndOfWarOut* signal).

Input: This model requires two dates from the user: the start of the war (*E-Time* block) and the end of the conflict (*Reconstitution* block). Three input pipelines are connected to this block: aircraft deployed to the FOL (*ACDeployedIn*), spare engines sent to the FOL (*DeployedEngineIn*), and engines transported back from an FSL (*FSLEngineIn*).

Output: No analysis tool has been designed for this block. However, the user may observe the number of aircraft moving in and out of

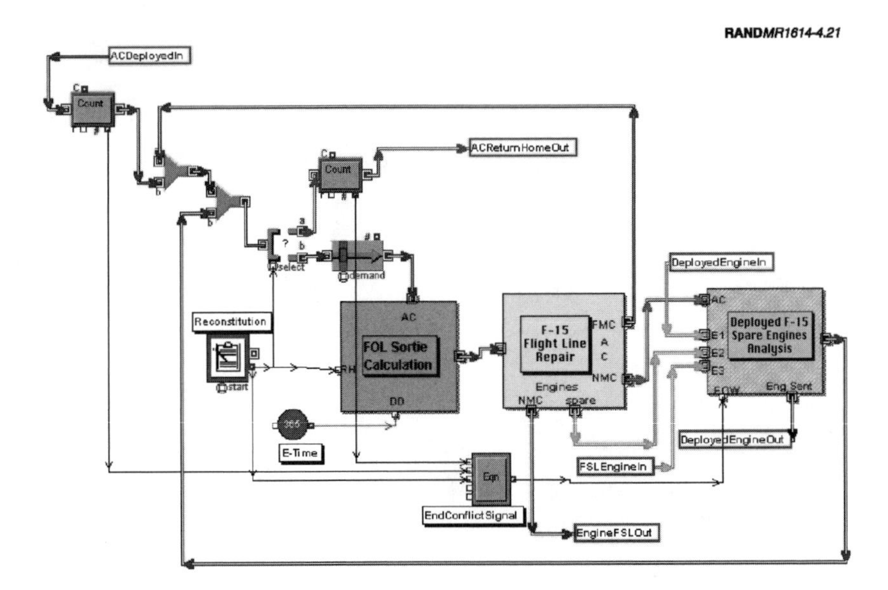

Figure 4.21—FOL Block

other blocks. There are three output pipelines: the deployed aircraft returning to their home base (*ACReturnHomeOut*), engines shipped to an FSL for repair (*EngineFSLOut*), and engines sent back to home bases after the conflict (*DeployedEngineOut*).

Chapter Five

CONCLUSIONS AND RECOMMENDATIONS

Given the nature of this report as a technical guide to the EnMasse simulation model, we again invite readers to refer to MR-1431-AF, *Supporting Expeditionary Aerospace Forces: Alternatives for Jet Engine Intermediate Maintenance,* for the detailed analysis of engine maintenance option policies and its resulting conclusions. We can, however, make certain conclusions and recommendations regarding building a model using such GUI-based software as Extend.

A number of simulation models treat repair systems and have been used for analyses similar to ours, the most notable of which is Dyna-METRIC, which was developed by RAND in the 1980s and used extensively in various versions for a number of studies during the 1980s and early 1990s. However, even the most recent versions of DynaMETRIC do not satisfy the requirements outlined above. They do not track individual units with specific properties and do not have much detail in their representation of transportation. Further, they are written in FORTRAN and difficult to modify internally to handle some of these potential extensions.

Many of these drawbacks are absent in current GUI-simulation packages. These packages, drawing on progress in programming languages, user interfaces, and hardware capabilities, make it possible to quickly design, write, and use simulations whose complexity and detail would have been impossible with the computing resources available only a decade ago.

Building EnMasse using Extend blocks allowed us to identify individual engines and aircraft and capture detailed information about their status and progress in events ranging from flying sorties to maintenance. Crucial management decisions in engine repair are based on these characteristics, and EnMasse allows decisionmakers to evaluate potential alternative maintenance policies, such as reliability-centered maintenance.

EnMasse is flexible enough to be used for further analysis in future expanded studies of engine repair that can incorporate other important characteristics, such as the management of engine deployment and repair based on the time characteristics of individual engines, the effects of engine demographics and different management decisions on JEIM and depot workload, more detailed representations of repair modes based on whether an engine removal is scheduled or unscheduled, and transportation policies. With EnMasse, these extensions can be easily and naturally added in the future.

Extend (and ultimately EnMasse) is somewhat limited in terms of input and output generation. Although we have centralized most of the input parameters in one place and Extend allows writing of output to external files, further work may be needed to make the model easier to use. A centralized external database where all possible input parameters could be read would greatly enhance EnMasse and allow for a better parametric analysis of the system. Although the current version of Extend software allows for basic sensitivity analysis, this feature is not flexible enough and will force additional "coding" for certain analyses of the engine maintenance system.

Appendix A

A SAMPLE RUN

In this appendix, we present some of the input and output parameters used in our analysis. We illustrate these parameters by running an FSL scenario for the F100-229 engines.

MODEL SETUP

This model has two F-16 bases with 18 Primary Aircraft Authorized (PAA) each and two F-15 bases with 18 and 48 PAA. The utilization rates for the F-16s and F-15s are set at 19 and 18, respectively. We deploy two-thirds of the aircraft to two single–Mission Design Series (MDS) forward operating locations (FOLs) with a single forward support location (FSL) serving both FOLs. The war starts on day 365 of the simulation with 10 days of surge and 90 days of sustained flying. The simulation terminates on day 730. Engines are removed at the rate of 3.5 per 1,000 flying hours for unscheduled engine removal (UER) and 1.5 per 1,000 flying hours for scheduled engine removal. We run the model for two simulated years.

Three rail teams are assigned to each F-16 base and the smaller F-15 base, and seven rail teams are assigned to the larger F-15 base. The larger F-15 base deploys three of its rail teams and the others deploy one each. Furthermore, two rail teams from other units augment the FSL, for a total of eight rail teams at the forward maintenance shop.

The current spares level and the war reserve engine (WRE) level are shown in Table A.1. Each unit has a single test cell and the FSL has two test cells.

RESULTS OF THE SIMULATION

The model generates aircraft tail number, engine demography, and serial number. Table A.2 presents sample engine demography for an F-16 base, in which age refers to the total age of the engine in hours and cycle is the total number of flying hours since the last flightline inspection.

On average, each F-16 base requires 12 FMC aircraft per day during peacetime and about four when the rest of the force is engaged at the FOL. The results for both the home units and the FOLs are reported in Table A.3.

For our sample run, no sorties were lost at the home units, but there were several missed sorties at the F-16 FOL.

Table A.1

Spares and WREs at the Units

	18 PAA F-16	18 PAA F-16	18 PAA F-15	48 PAA F-15
Spares	4	10	12	24
WREs	1	3	3	8

Table A.2

F-16 Engine Demography

Arrival (Days)	Priority	Age (Hours)	Cycle (Hours)	Engine Number	Tail Number
176	297.5	297.5	109.5	1115	6498
176	298.0	298.0	138.0	1234	9038
176	298.0	298.0	165.0	1035	8967
176	298.0	298.0	177.0	4380	5016
176	300.5	300.5	130.5	1948	7895
177	285.0	285.0	135.0	1621	7341
177	289.5	289.5	169.5	1780	5665
177	299.0	299.0	111.0	1115	6498
178	230.5	230.5	121.5	1399	6882
178	299.5	299.5	139.5	1234	9038
178	299.5	299.5	166.5	1035	8967
178	299.5	299.5	178.5	4380	5016

Figure A.1 illustrates two different simulation runs[1] for this FOL showing as high as 17 percent sortie loss (or about eight missed sorties out of 44) on day 374 (nine days into the conflict).

Table A.3

Aircraft Requirements at the Units and the FOLs

	F-16	F-15
Units	24, 8	40, 14
FOLs	23, 12	35, 22

NOTE: Unit numbers are peacetime aircraft requirements and aircraft requirements while the rest of the force is engaged. FOL numbers are surge and sustained aircraft requirements.

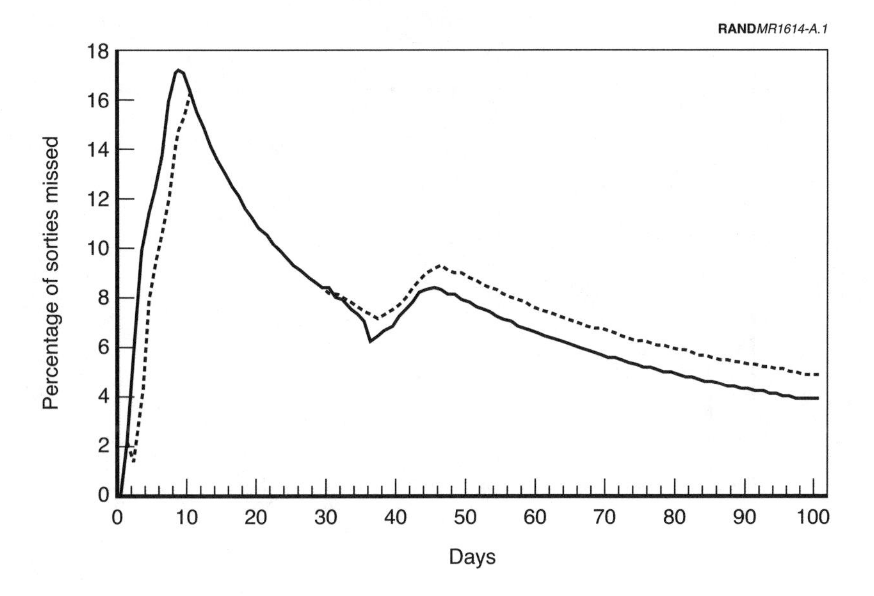

Figure A.1—Percentage of Missed Sorties at the F-16 FOL

[1]The data are generated by the *Sortie Shortfall* block (see Figure 4.6).

The JEIM Shop

The JEIM shop with its eight rail teams maintained an average of 35 scheduled and 135 unscheduled engines during the conflict, with 41 engines incurring an additional average delay of about 9.6 days stemming from engine not mission capable because of supply (ENMCS).[2] Figure A.2 illustrates the daily number of engines at the JEIM shop during this period.

Figure A.3 illustrates the distribution of ENMCS delays per engine during the conflict,[3] where the engaged forces are given priority in the allocation of parts. Cumulatively, about 30 percent of all engines are delayed because of the lack of appropriate parts, including modules. As the figure indicates, the majority of engines are in ENMCS

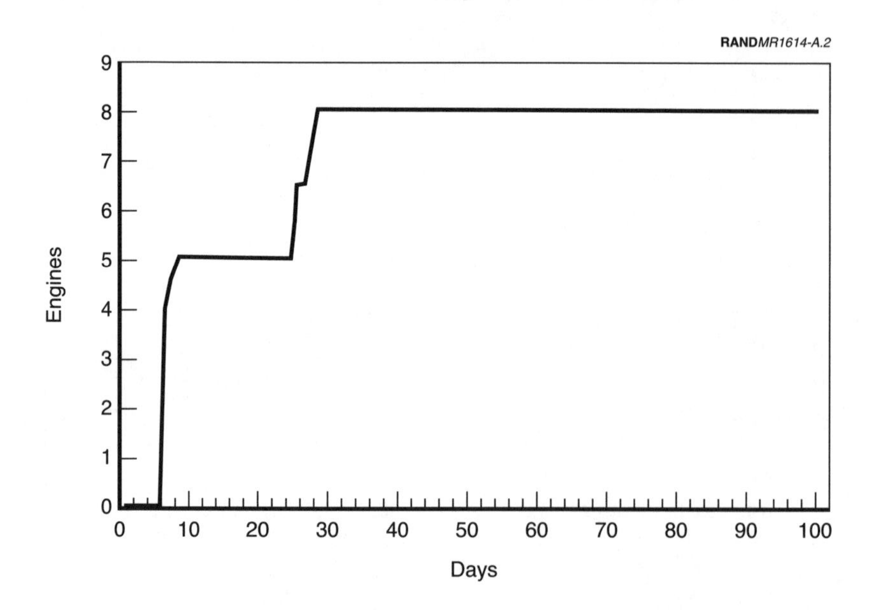

Figure A.2—Daily Number of Engines at the JEIM Shop

---

[2]This figure is produced from the *JEIM Shop* block (see Figure 4.8).

[3]Data are from the random number generator (*Rand*) in the *AWP* block (see Figure 4.9).

for just a few days, but a small portion of them are delayed as long as 68 days.

Test Cell Performance

One of the key resources in the model is the hush house at the JEIM shop. Figure A.4 illustrates the utilization of the test cells at the forward support hush house. In this model, we allocated two test cells to the FSL and, as the figure indicates, both test cells were needed to meet the demand. In fact, the average wait time for a test cell was about five hours and the longest queue length was about three. To illustrate the importance of the number of test cells, we ran the model using only one test cell. The consequences were rather devastating for the FOLs because 30 percent of the engines queued for the test cell were never serviced and therefore the FOLs missed up to 8 percent of their sorties. The average wait time of the test cell grew to 12 days, with some engines waiting as long as 26 days for the hush house.

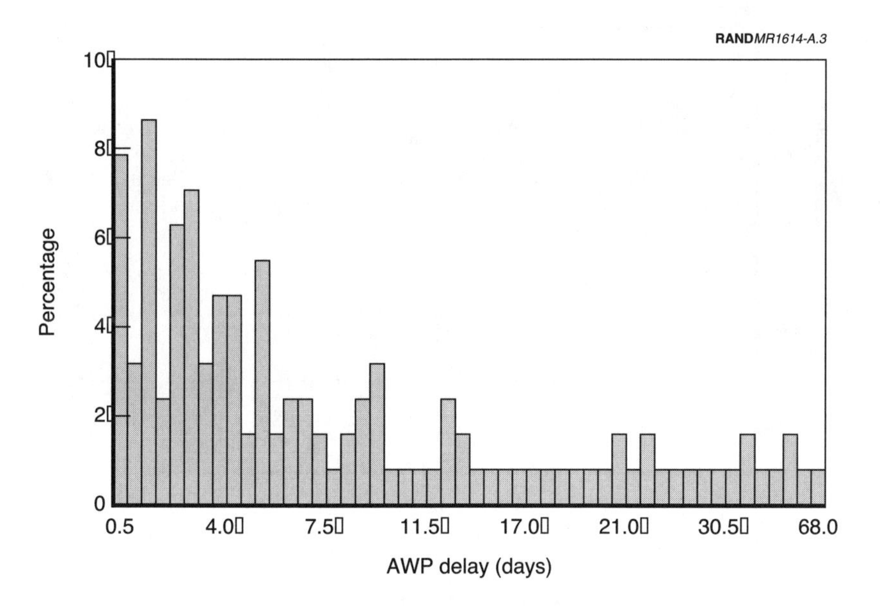

Figure A.3—ENMCS Distribution

Spare Engines Analysis

One of the key performance measures is the availability of spares at the operating locations. Figure A.5 illustrates the typical dynamics of serviceable spares at a single base.[4] Aircraft quickly absorb the initial number of spares (i.e., spares allocated to the base at the beginning of the simulation) in the first few weeks, followed by a series of recoveries and losses. At the start of the conflict (period highlighted starting at day 360), the number of spares dramatically drops as engines are deployed to forward locations. However, after the conflict, a quick but small recovery occurs, followed by a series of recoveries and losses.

At the FOLs, the dynamics are somewhat different because the FSL requires only a few days to become fully operational and the number and duration of the sorties increase significantly. Figure A.6 illustrates the daily serviceable spares at the F-15 FOL with the standard

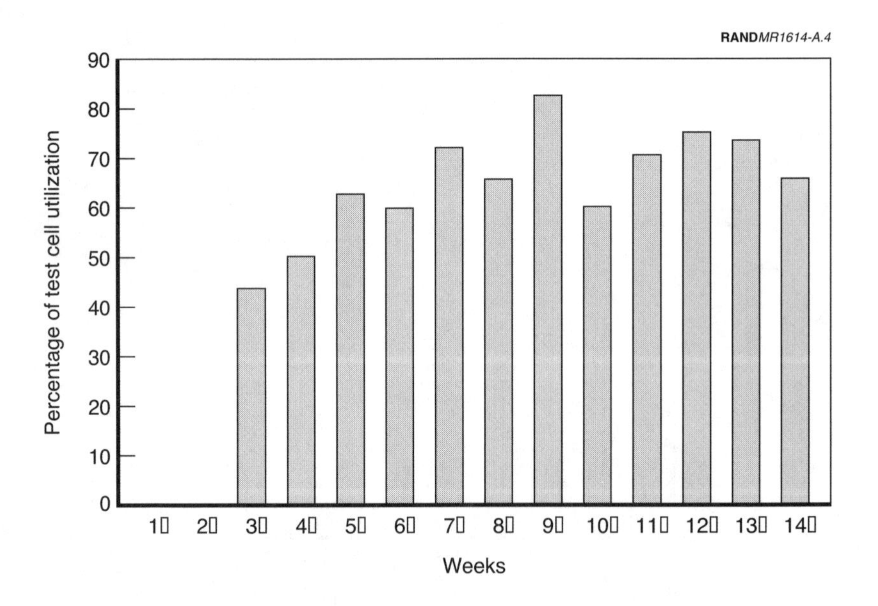

Figure A.4—Weekly Test Cell Utilization at the FSL

[4]Data are from the *Spares Analysis* block of an F-15 base (see Figure 4.16).

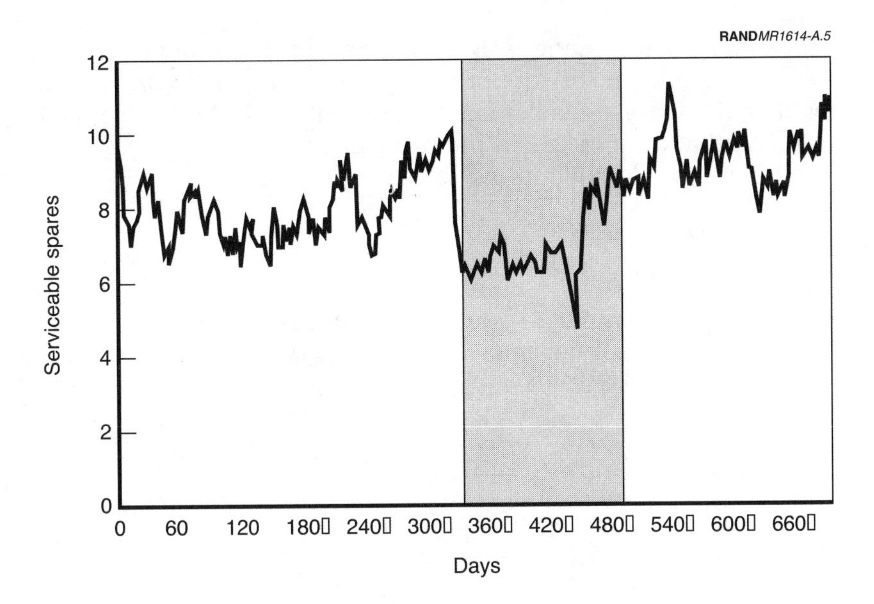

Figure A.5—Serviceable Spares at an F-15 Base

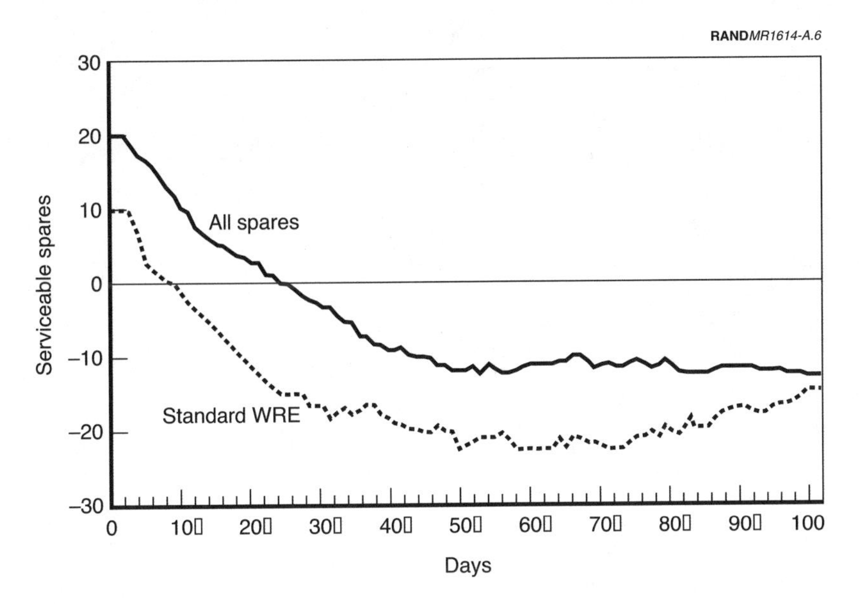

Figure A.6—Serviceable Spares at the F-15 FOL

WRE and the deployment of all available spares.[5] It is not surprising that the increase in the number of initial spares has improved the health of the spare levels throughout the conflict. It should be noted, however, that an increase in the number of rail teams beyond the current level would not improve the spares levels.[6]

---

[5]Data are from the *Spares Analysis* block of an F-15 FOL (see Figure 4.16).

[6]There is a diminishing return for a marginal increase in the number of rail teams. The current rail team allocation is optimal for this FSL.

Appendix B

GLOSSARY OF EnMasse FIGURES AND ICONS

This appendix is a list of EnMasse blocks as they appear in the library. We start with the most basic blocks and build toward the uppermost hierarchy of the EnMasse library. Most of the substantive blocks from EnMasse have been discussed in detail in the text, and so detailed input and output information is not repeated. For those not discussed, their function, the discussion of the other blocks, and the structure of the real-world repair system should suffice to identify inputs and outputs.

EnMasse BASIC BLOCKS

This section (Figures B.1 through B.12) lists the most basic blocks of the EnMasse library. Note that some of the less-common Extend blocks may be used.

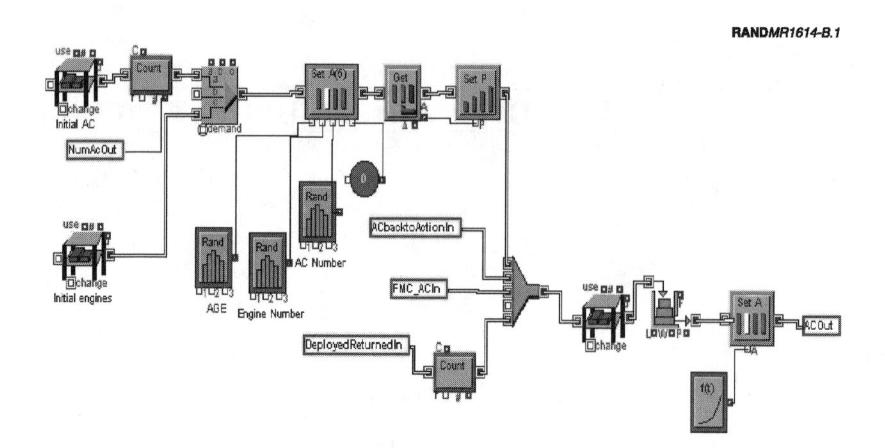

Figure B.1—Aircraft Engine Selection

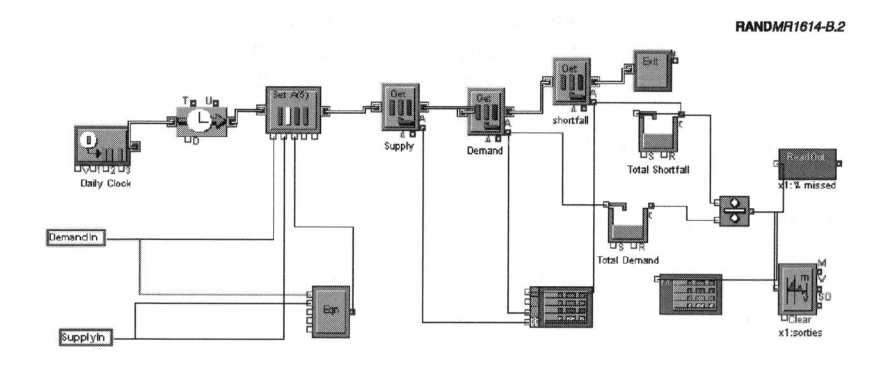

Figure B.2—Sortie Shortfall Block

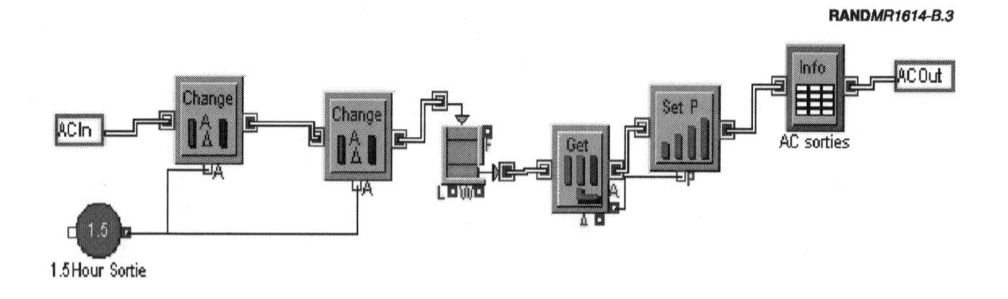

Figure B.3—Flight Sorties Block

RANDMR1614-B.4

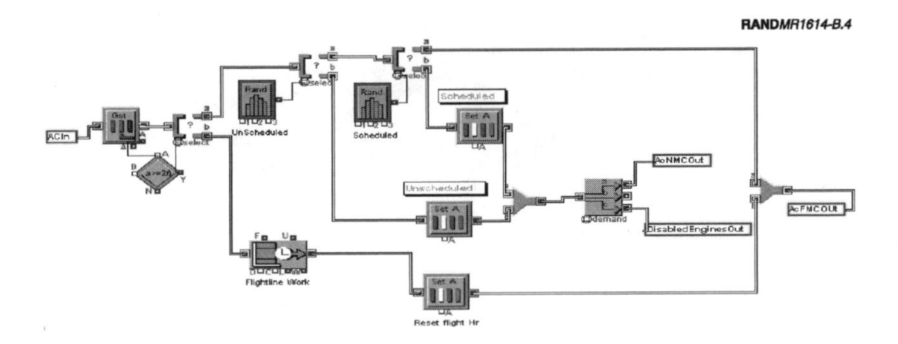

Figure B.4—Flightline Inspection Block

RANDMR1614-B.5

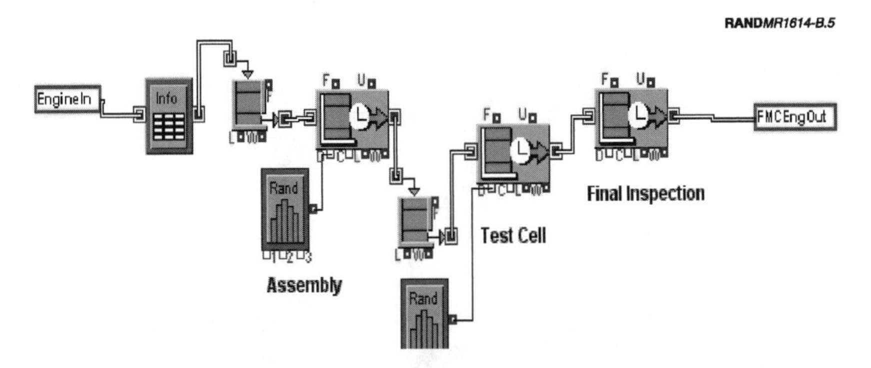

Figure B.5—Assembly and Test Cell Block

RANDMR1614-B.6

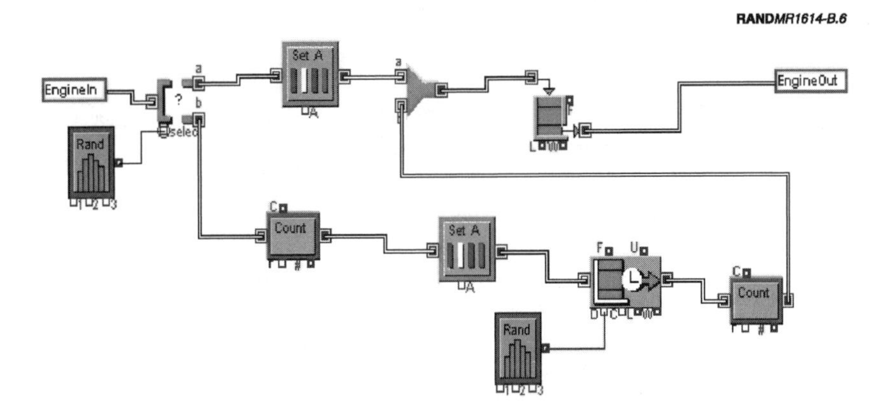

Figure B.6—AWP Block

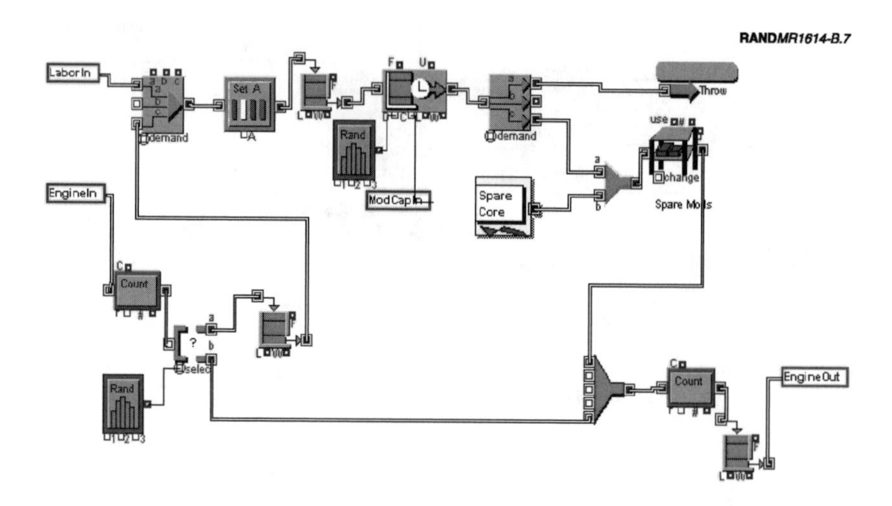

Figure B.7—Individual Module Repair

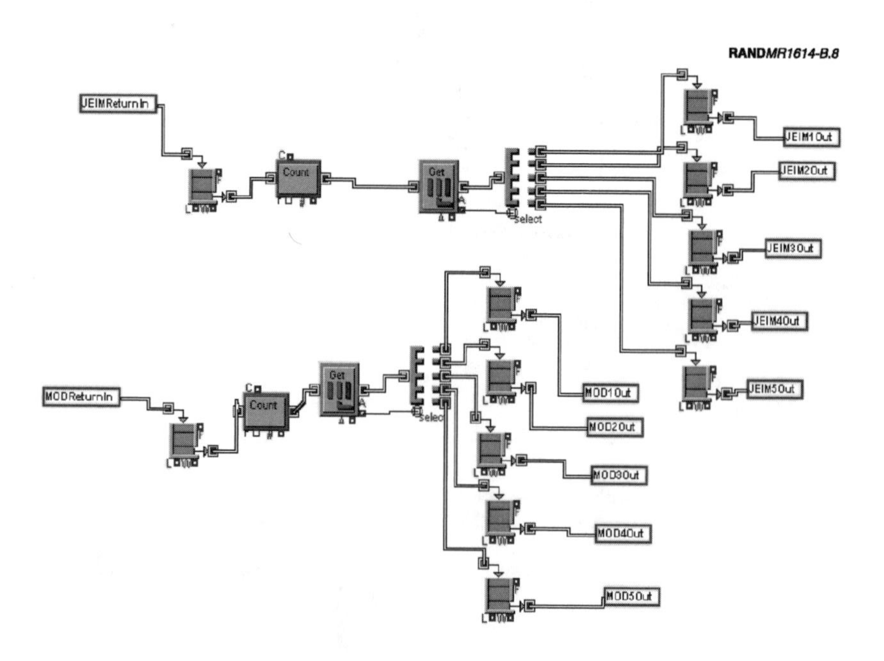

Figure B.8—Labor Reconstitution Block

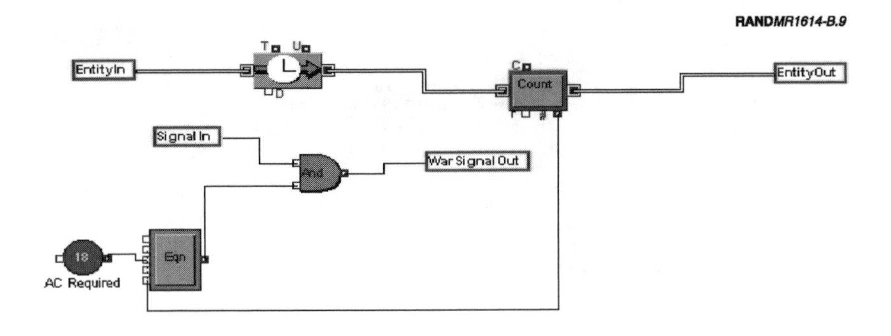

Figure B.9—Deployment Calculation Block

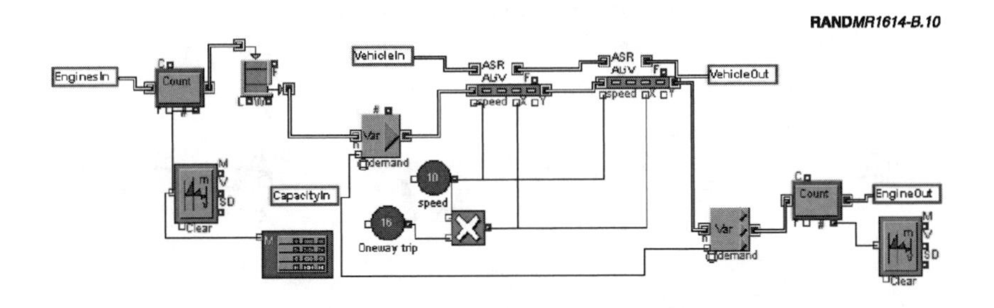

B.10—Transportation Block

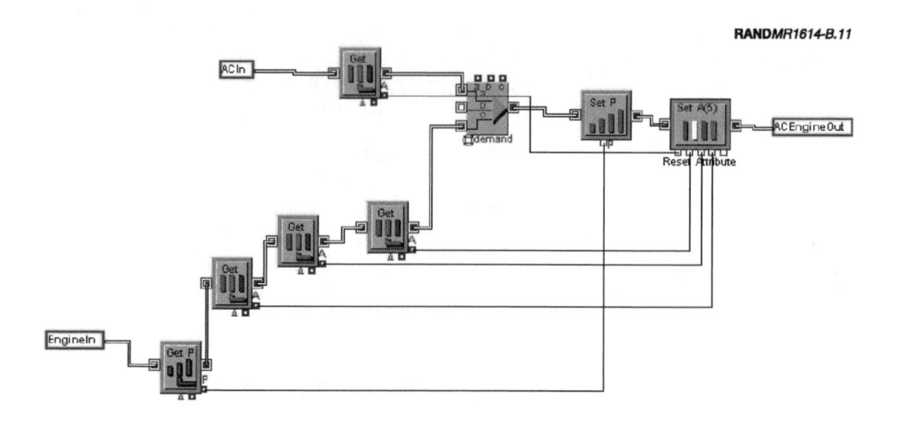

Figure B.11—Reset Attribute Block

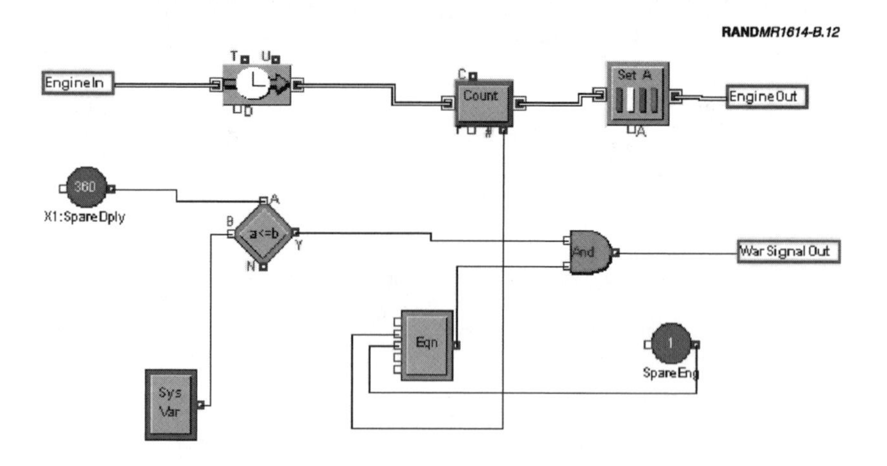

Figure B.12—Spare Engines (WRE) Deployment Block

HIGHER-LEVEL BLOCKS

In this section (Figures B.13 through B.23), we present the next level of the hierarchy by combining the basic blocks of the EnMasse library.

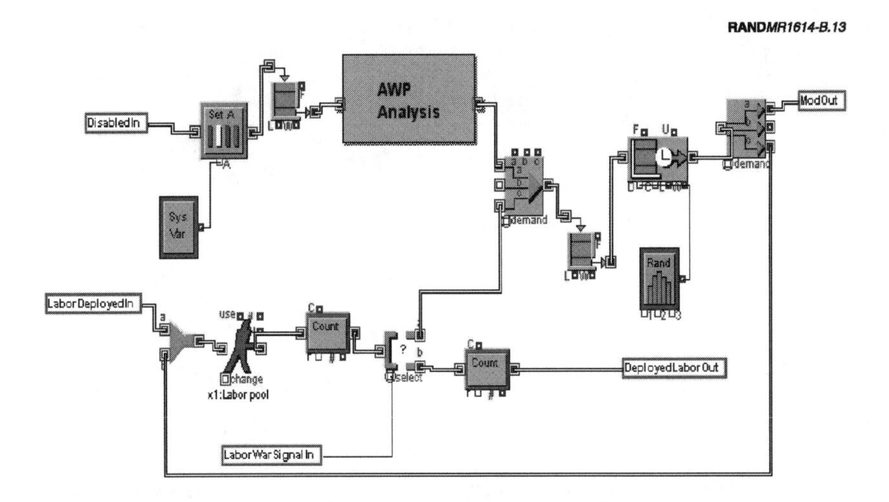

Figure B.13—JEIM Block at a Unit

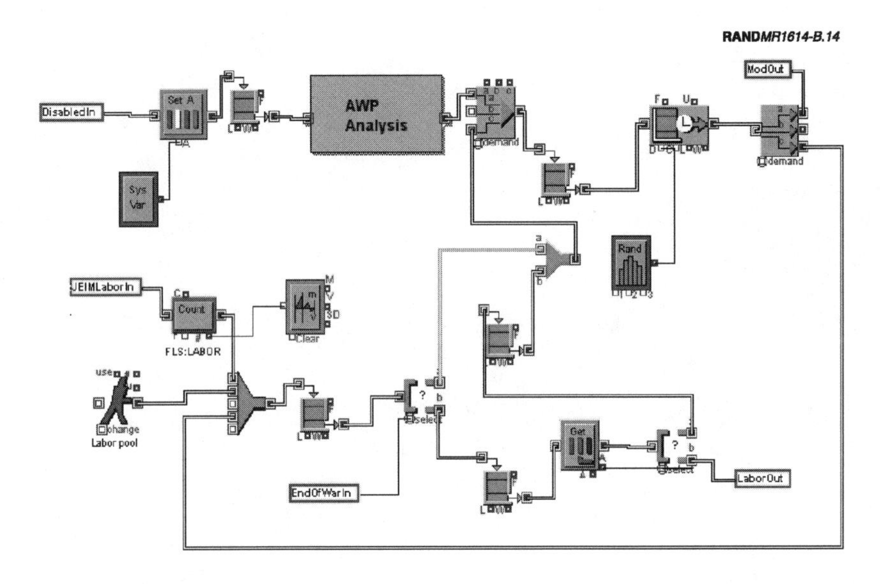

Figure B.14—JEIM Block in a Deployed Location (FSL or Deployed JEIM)

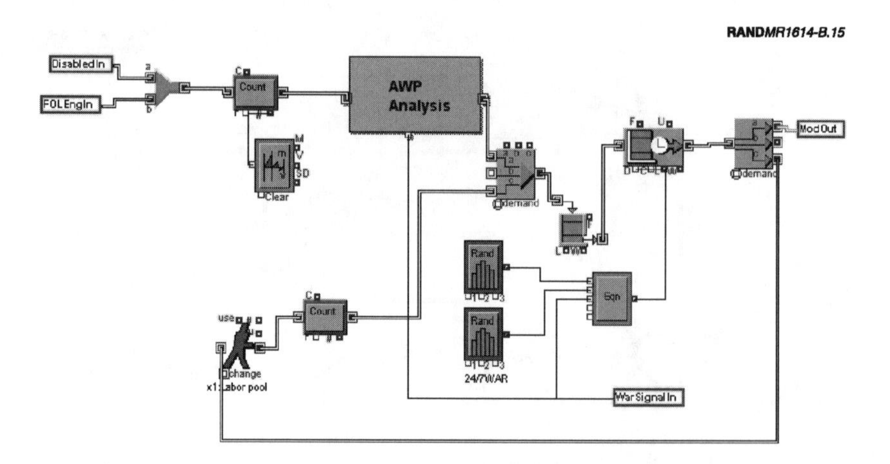

Figure B.15—JEIM Supporting Training and Deployed Units Block (CSL and Home Support)

RANDMR1614-B.16

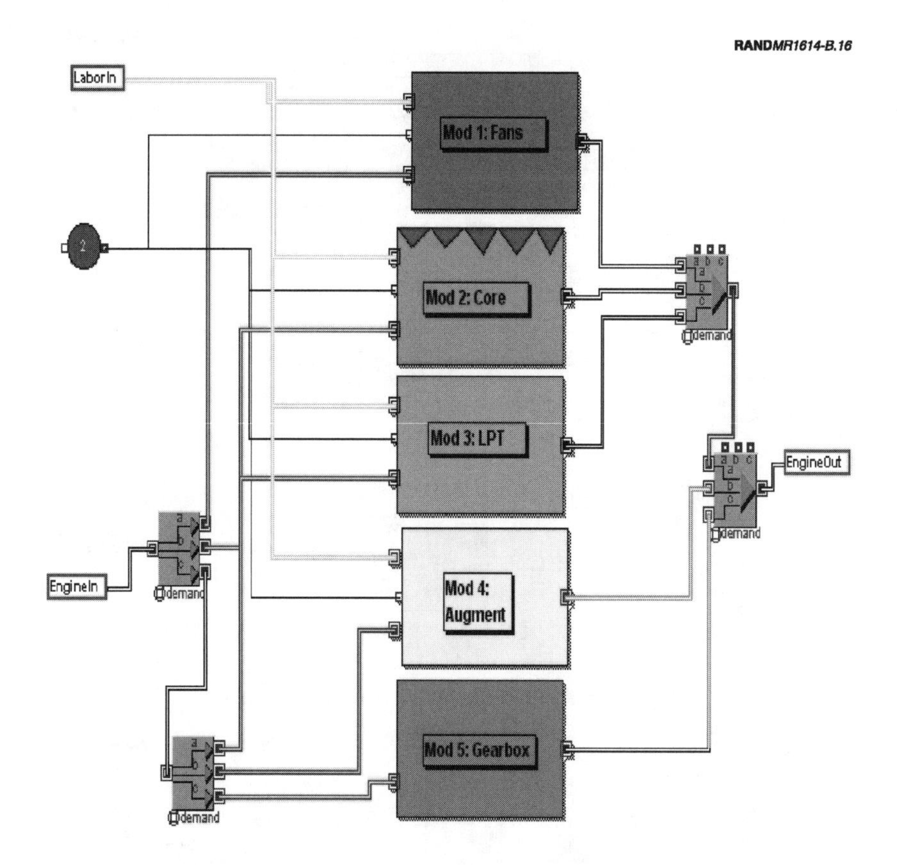

Figure B.16—Module Repair Block

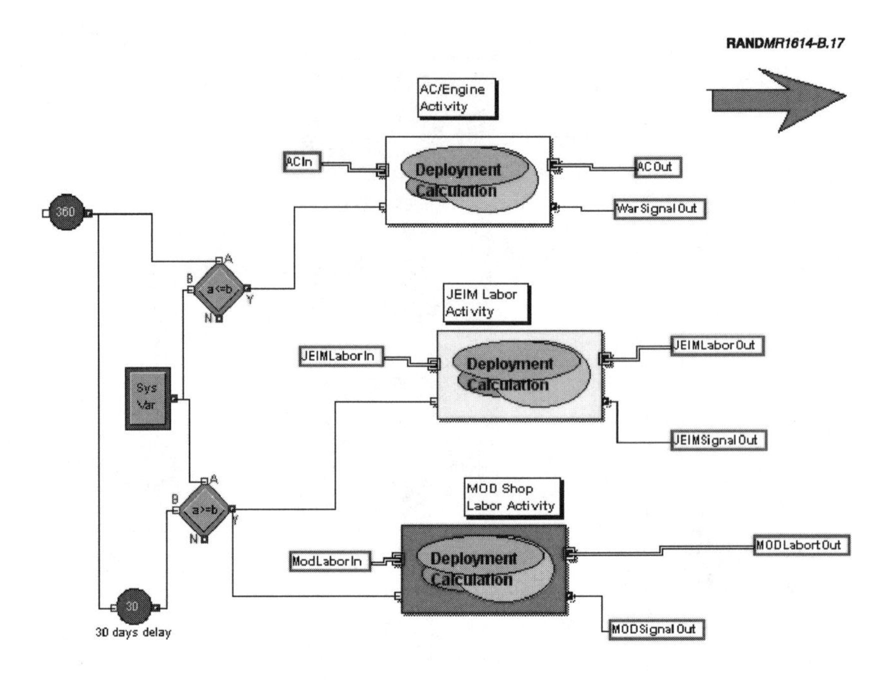

Figure B.17—Resource Computation Block

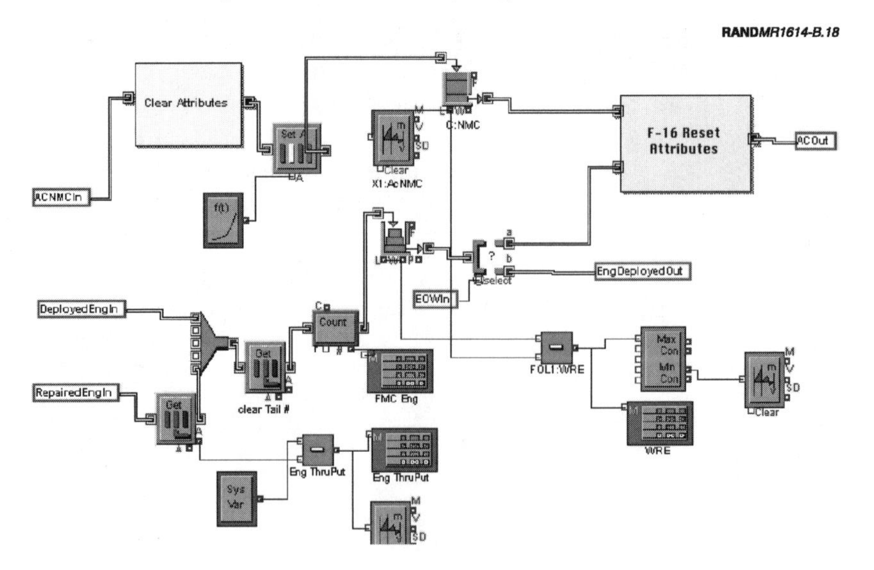

Figure B.18—Spare Engines Analysis Block

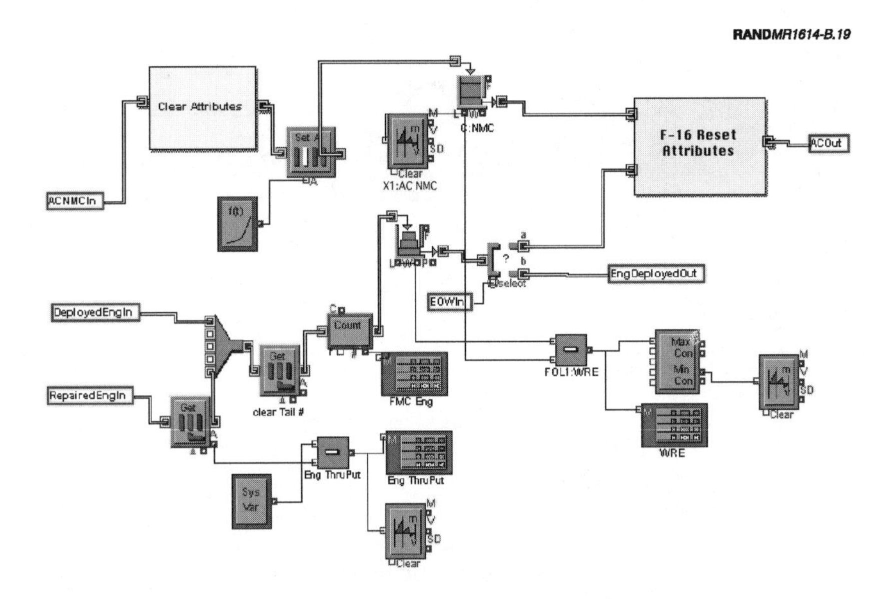

Figure B.19—Deployed Location Spare Engines Analysis Block

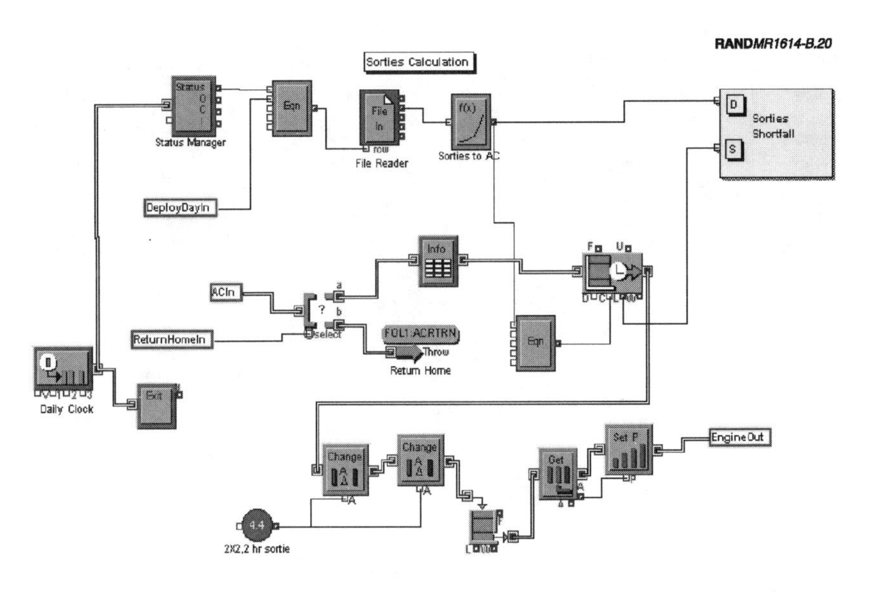

Figure B.20—FOL Sortie Calculation Block

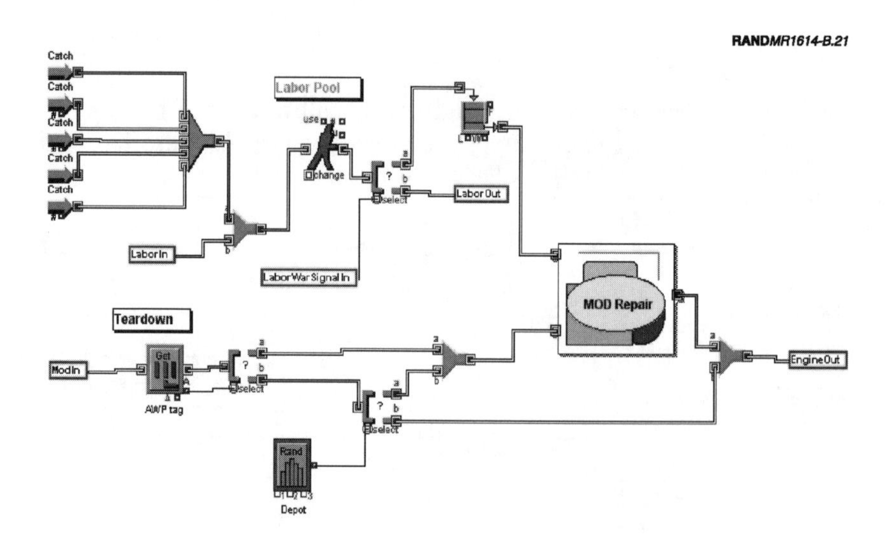

Figure B.21—Module Shop Block at the Units

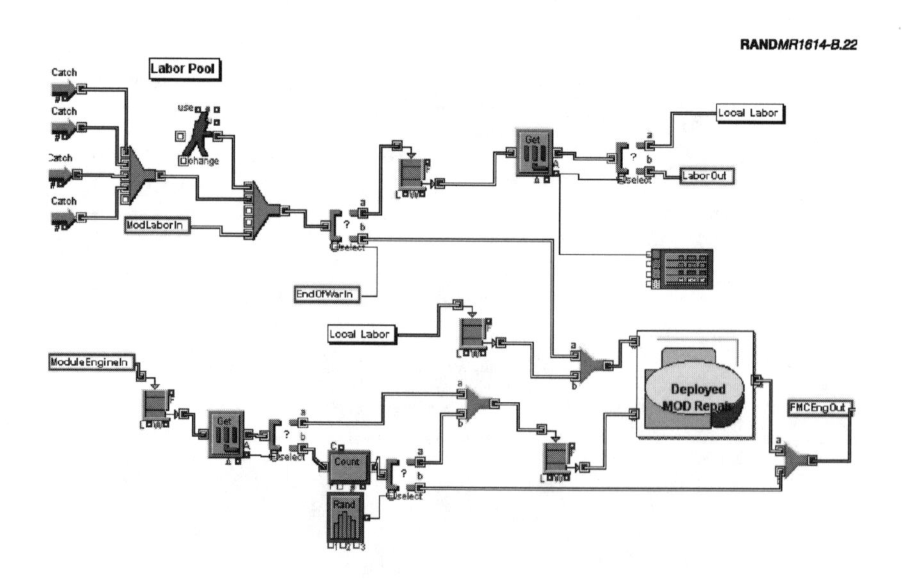

Figure B.22—Deployed Module Shop Block (FSL, Deployed JEIM)

UPPER-LEVEL BLOCKS

This section (Figures B.24 through B.27) will cover the rest of the blocks in the EnMasse library. At the next level of the hierarchy, we can represent bases, FOLs, FSLs, and CSLs.

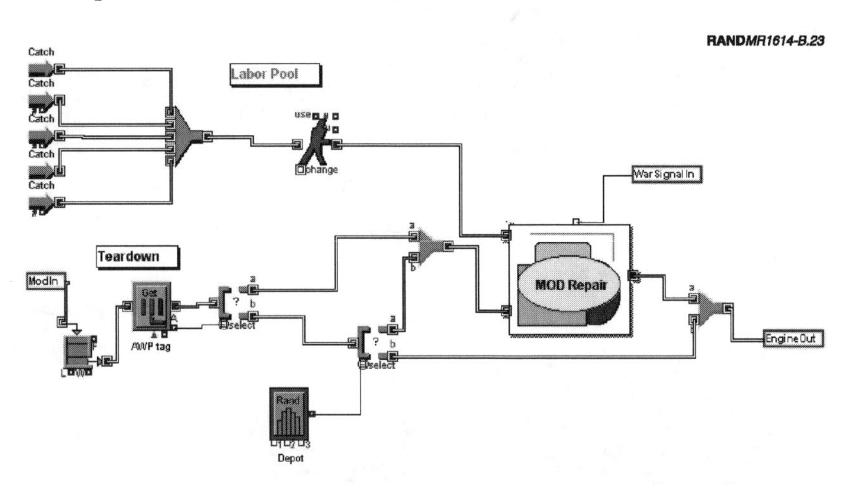

Figure B.23—Module Shops Supporting Training and Deployed Units (CSL, Home Support)

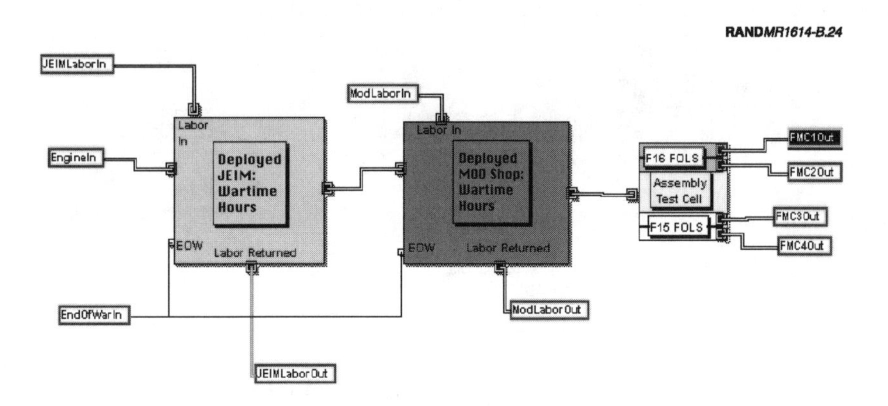

Figure B.24—Engine Maintenance at an FSL or an FOL

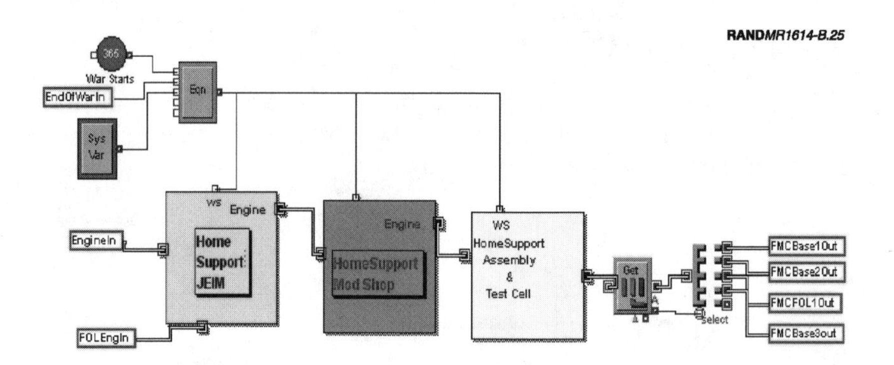

Figure B.25—Engine Maintenance at a CSL

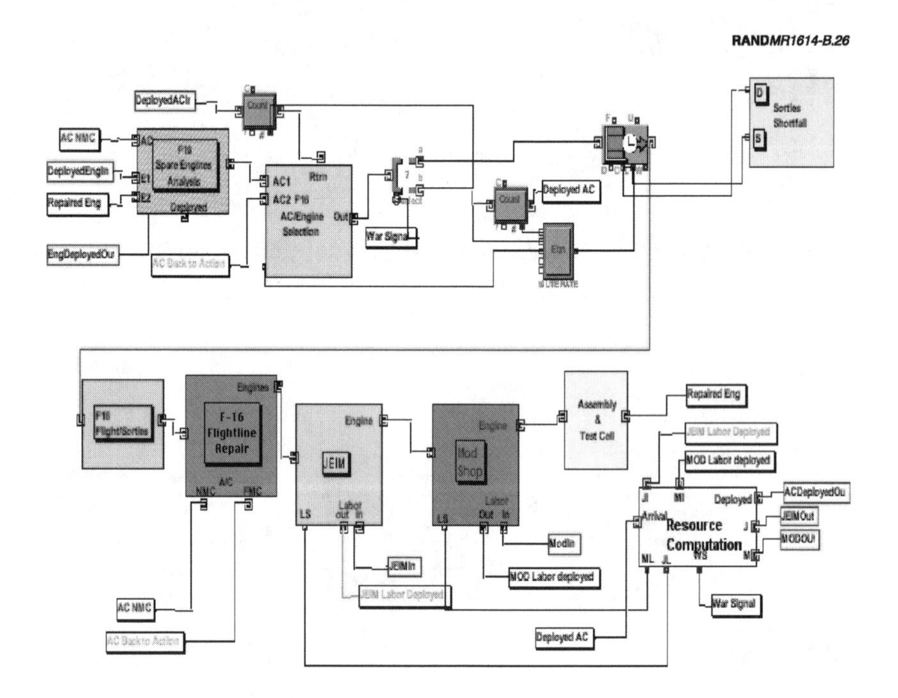

Figure B.26—MDS-Based Unit Block

RANDMR1614-B.27

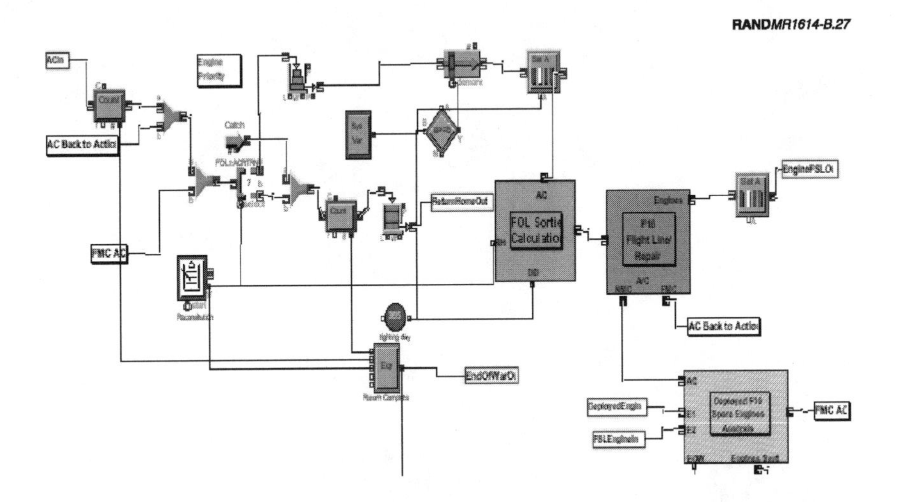

Figure B.27—FOL Block

BIBLIOGRAPHY

Amouzegar, Mahyar, Lionel A. Galway, and Amanda Geller, *Supporting Expeditionary Aerospace Forces: Alternatives for Jet Engine Intermediate Maintenance,* Santa Monica, Calif.: RAND, MR-1431-AF, 2002.

Davis, Richard G., *Immediate Reach, Immediate Power: The Air Expeditionary Force and American Power Projection in the Post Cold War Era,* Washington, D.C.: Air Force History and Museums Program, 1998.

Feinberg, Amatzia, Hyman L. Shulman, Louis W. Miller, and Robert S. Tripp, *Supporting Expeditionary Aerospace Forces: Expanded Analysis of LANTIRN Options*, Santa Monica, Calif.: RAND, MR-1225-AF, 2001.

Galway, Lionel A., Robert S. Tripp, Timothy L. Ramey, and John G. Drew, *Supporting Expeditionary Aerospace Forces: New Agile Combat Support Postures*, Santa Monica, Calif.: RAND, MR-1075-AF, 2000.

Isaacson, Karen, and Patricia Boren, *Dyna-METRIC Version 6: An Advanced Capability Assessment Model*, Santa Monica, Calif.: RAND, R-4214-AF, 1993.

Killingsworth, Paul S., Lionel A. Galway, Eiichi Kamiya, Brian Nichiporuk, Timothy L. Ramey, Robert S. Tripp, and James C. Wendt, *Flexbasing: Achieving Global Presence for Expeditionary Aerospace Forces,* Santa Monica, Calif.: RAND, MR-1113-AF, 2000.

Peltz, Eric, Hyman L. Shulman, Robert S. Tripp, Timothy Ramey, Randy King, and John G. Drew, *Supporting Expeditionary Aerospace Forces: An Analysis of F-15 Avionics Options,* Santa Monica, Calif.: RAND, MR-1174-AF, 2000.

Ryan, General Michael E., *Evolving to an Expeditionary Aerospace Force,* Commander's NOTAM 98-4, Washington, D.C., July 28, 1998.

Tripp, Robert S., Lionel A. Galway, Paul S. Killingsworth, Eric Peltz, Timothy L. Ramey, and John G. Drew, *Supporting Expeditionary Aerospace Forces: An Integrated Strategic Agile Combat Support Planning Framework,* Santa Monica, Calif.: RAND, MR-1056-AF, 1999.

Tripp, Robert S., Lionel A. Galway, Timothy L. Ramey, and Mahyar Amouzegar, *Supporting Expeditionary Aerospace Forces: A Concept for Evolving the Agile Combat Support/Mobility System of the Future,* Santa Monica, Calif.: RAND, MR-1179-AF, 2000.